THE EDGE OF PROGRESS

When ambition shapes the world, the edge is where stories unfold

The Edge of Progress

ISBN 978-631-00-8282-0

Print Edition

All content, including illustrations, text, layout, and design, was created by Rocio Luz Fernandez.
Photographs at the end of each comic are by the author, unless otherwise credited.

Dedicated to those who forge their own miracles—
and to Facundo and his hero: his dad, Gaston.

WHERE STORIES UNFOLD

BEFORE WE BEGIN

They said it couldn't be done—too long, too high, too dangerous. But what is the edge of progress, if not the challenge of defying the impossible?

Through sheer determination, engineers and visionaries shattered limits—carving canals across continents, bridging impossible spans, and taming nature with towering dams. This documentary comic, *The Edge of Progress*, chronicles five historic feats that redefined what was possible in their time—and still stand as monuments to human ambition:

The Longest Suspension Bridge (*United States*, 1883)
The Brooklyn Bridge connected two great cities with a span **50% longer** than its nearest rival. Its towers—then the tallest ever built for a bridge—have long stood as monuments to vision, sacrifice, and resilience.

The Tallest Dam (*Italy*, 1960)
The Vajont Dam was a bold attempt to harness nature's power, rising **10% higher** than any dam ever built up to that point.

The Largest Man-made Lake (*Argentina*, 1891): The San Roque Dam created a reservoir **80% larger** than any that had come before—an audacious response to surging demand and political firestorms.

The Longest Freshwater Pipeline (*Australia*, 1903)
The Goldfields Water Supply Scheme set a record for the longest steel pipeline built to carry water—and hope—to Australia's arid interior.

The First Interoceanic Canal (*Panama*, 1914)
The Panama Canal connected oceans and continents through relentless determination and innovation. A marvel of its time—its impact remains unmatched.

Each of these projects faced formidable challenges: harsh climates, political struggles, financial obstacles, and the unyielding force of nature. Some triumphed, forever changing the course of history. Others came at a tremendous cost.

What does it take to build what has never been built before?
What does it feel like to walk along the edge of progress?

Step into the past and witness the remarkable feats of those who dared to envision—and build—the "impossible".

The East River, a formidable divide between Manhattan and Brooklyn, had long stood as an obstacle to progress. Engineers dreamed of a bridge long enough to span its vast waters, linking two of the fastest-growing cities in the world.

But this would be no simple feat. The bridge had never been built before, and the very idea of such a structure seemed impossible. It would take not only ingenuity and perseverance but a family willing to risk everything to make it a reality—against all odds and at great personal cost.

This is the story of the Brooklyn Bridge.

- 1 -

SUSPENDED DREAMS

A Bridge to the Future, Forged in Perseverance

IN 1865, NEW YORK AS WE KNOW IT DIDN'T EXIST YET. IT WAS A COLLECTION OF SEPARATE CITIES—EACH WITH ITS OWN IDENTITY, HISTORY, AND AMBITIONS.
AT 92 METERS TALL, TRINITY CHURCH'S STEEPLE DOMINATED THE SKYLINE—THE CITY'S TALLEST STRUCTURE.
THE BRONX
QUEENS
East River
Houdson River
MANHATTAN
BROOKLYN
FULTON FERRY
JERSEY FERRY
STATEN ISLAND
BROOKLYN FERRY
WITH THE CIVIL WAR OVER, INDUSTRY AND IMMIGRATION BOOMED. MANHATTAN AND BROOKLYN GREW FAST—TWO OF THE BUSIEST CITIES IN THE NATION.
THE EAST RIVER WAS A LIFELINE AND A BARRIER. OVER 70 MILLION PEOPLE CROSSED THE RIVER EACH YEAR ABOARD FERRY LINES—HALF ABOARD "UNION FERRY", THE BUSIEST SERVICE IN THE WORLD.
FOR 70 YEARS, BRIDGES OVER THE EAST RIVER REMAINED A DREAM—DISMISSED AGAIN AND AGAIN. ANY BRIDGE HAD TO SPAN A DAUNTING 1,825-METER GAP, WITHSTAND STRONG CURRENTS, AND ANCHOR DEEP INTO THE RIVER'S UNSTABLE, MUDDY BOTTOM.
THE MUDDY RIVERBED NEEDS DEEPER FOUNDATIONS. BEDROCK IS AT 44 METERS.
WHO COULD BELIEVE A BRIDGE IS POSSIBLE?
ANY BRIDGE WOULD ALSO NEED TO CLEAR ENOUGH SPACE FOR TALL-MASTED SHIPS TO PASS UNDERNEATH.

NO BRIDGE OF THIS SCALE HAD EVER BEEN BUILT. TO SUCCEED, IT HAD TO BE THE LONGEST AND TALLEST SUSPENSION BRIDGE THE WORLD HAD SEEN!
IT SHOULD BE A SUSPENSION BRIDGE! THEY USE FEWER PIERS—PERFECT FOR MUDDY RIVERBEDS.
SUSPENSION BRIDGES ALREADY EXISTED, BUT NONE HAD EVER FACED THE EAST RIVER'S RELENTLESS CURRENTS—SURGING AT EIGHT KNOTS.
BUT THIS GAP... IT'S IMPOSSIBLE FOR ANY KIND OF BRIDGE!
THEY ALSO MAXIMIZE CLEARANCE, PERFECT FOR RIVER TRAFFIC.

THEN CAME THE BRUTAL WINTER OF 1866–67. THE EAST RIVER FROZE SOLID, CUTTING OFF FERRY TRAFFIC.
THE FERRY'S STUCK! WE NEED A BRIDGE—NOW!
THIS IS UNBEARABLE! WE NEED A BRIDGE NOW!
WE NEED THAT BRIDGE NOW!
FRUSTRATION BOILED OVER—THE PEOPLE DEMANDED A BRIDGE! BUT WHO HAD THE VISION—AND THE DARING—TO MAKE IT HAPPEN?

THE ANSWER? JOHN AUGUSTUS ROEBLING, A PRUSSIAN IMMIGRANT, INVENTOR, AND ENGINEERING GENIUS WHO ARRIVED IN AMERICA IN 1831 SEEKING FORTUNE.
J. A. ROEBLING SUSPENSION BRIDGE (IN CINCINNATI)
ROEBLING WAS OBSESSED WITH SUSPENSION BRIDGES. HE HAD ALREADY DESIGNED SOME OF THE MOST REMARKABLE IN THE U.S., BUT THIS? THIS WOULD BE HIS MASTERPIECE!
DELAWARE AQUEDUCT (IN LACKAWAXEN)

IN 1845, ROEBLING HAD PATENTED STRONGER, MORE FLEXIBLE STEEL CABLES—AN INNOVATION USED IN AQUEDUCTS, MINES, AND MORE... AND ONE THAT MADE HIM A FORTUNE.
STRONG YET LIGHTWEIGHT, ROEBLING'S CABLES MADE LONG-SPAN SUSPENSION BRIDGES BEGIN TO LOOK POSSIBLE—CAPABLE OF CARRYING MASSIVE LOADS ACROSS GREAT DISTANCES.
IN 1867, ROEBLING UNVEILED HIS VISION: A BRIDGE 50% LONGER THAN ANY BEFORE AND THE FIRST TO BE SUSPENDED BY STEEL CABLES.
"WHEN BUILT, IT WILL BE THE GREATEST ENGINEERING WORK ON THE CONTINENT, WITH TOWERS THAT WILL STAND AS NATIONAL MONUMENTS".
TOWERS WILL RISE LIKE A GOTHIC CATHEDRAL, A BLEND OF ART AND ENGINEERING.
TO ROEBLING, THIS WAS MORE THAN JUST A CROSSING BRIDGE—IT WAS A STRUCTURE THAT WOULD TRANSFORM TRADE, URBAN LIFE, AND THE WAY PEOPLE MOVED BETWEEN MANHATTAN AND BROOKLYN.
THAT'S WHY HE INTRODUCED A BOLD NEW IDEA: A PEDESTRIAN PROMENADE—SO PEOPLE COULD ENJOY THE BRIDGE, NOT JUST CROSS IT.
RAIL AND CARRIAGE LANES RAN ALONGSIDE SWEEPING RIVER VIEWS.

(*) ASCE: American Society of Civil Engineers.

AFTER STUDYING PNEUMATIC FOUNDATIONS IN EUROPE, HE WAS APPOINTED CHIEF ENGINEER OF THE BROOKLYN BRIDGE ON AUGUST 3, 1869.
"I AM THE ONLY LIVING MAN WHO HAD THE PRACTICAL EXPERIENCE TO BUILD THOSE GREAT CABLES."
I AM EMILY ROEBLING, HIS WIFE, AND I'M HERE TO HELP.
CONSTRUCTION OFFICIALLY BEGAN ON JANUARY 3, 1870.
"I KNOW THAT A SUCCESSFUL PROJECT WOULD HONOR MY FATHER, BUT FAILURE... THAT WOULD FALL ON MY SHOULDERS ALONE."
WASHINGTON HAD A CLEAR VISION, BUT THE OBSTACLES AHEAD WERE DAUNTING—HE WOULD HAVE TO OVERCOME CHALLENGES IN DESIGN, TECHNOLOGY, AND LABOR.
HIS WIFE, EMILY WARREN ROEBLING, WAS MORE THAN JUST A SUPPORTIVE PARTNER—SHE BECAME A STUDENT OF ENGINEERING.
THIS IS A ENORMOUS TASK... I MUST UNDERSTAND IT TO HELP WASHINGTON.
DURING THE BRIDGE'S EARLY STAGES, EMILY IMMERSED HERSELF IN TECHNICAL TEXTS, MEMORIZED COMPLEX DRAWINGS, AND MASTERED THE PRINCIPLES BEHIND HER HUSBAND'S AMBITIOUS DESIGN.

WASHINGTON DESIGNED THE BRIDGE'S FOUNDATIONS: TWO MASSIVE, AIRTIGHT YELLOW PINE CAISSONS, OPEN AT THE BOTTOM.
EACH TOWER ROSE FROM ITS OWN CAISSON—ONE ON EITHER SIDE OF THE RIVER.
WORKERS EXCAVATED THE RIVERBED BENEATH THE CAISSONS.
AS THEY DUG DEEPER, THE CAISSONS SANK GRADUALLY UNTIL THEY REACHED BEDROCK.
ONCE IN PLACE, THE CAISSONS WERE FILLED WITH CEMENT, FORMING A SOLID BASE FOR THE TOWERS ABOVE.
MUD TO EXCAVATE
ROCK TO REACH
INSIDE THE CAISSONS, AIR WAS PRESSURIZED TO KEEP OUT WATER AND MUD. WHEN WORKERS EMERGED TOO QUICKLY, THE RAPID CHANGE IN PRESSURE CAUSED AGONIZING PAIN, PARALYSIS, AND EVEN DEATH ("CAISSON DISEASE"—TODAY, WE KNOW IT AS DECOMPRESSION SICKNESS).
MORE THAN 110 WORKERS FELL ILL WITH CAISSON DISEASE—MANY WERE LEFT PERMANENTLY DISABLED.
IN MAY 1872, AFTER WORKING INSIDE A CAISSON, WASHINGTON COLLAPSED UPON REACHING THE SURFACE.
I FEEL DIZZY.

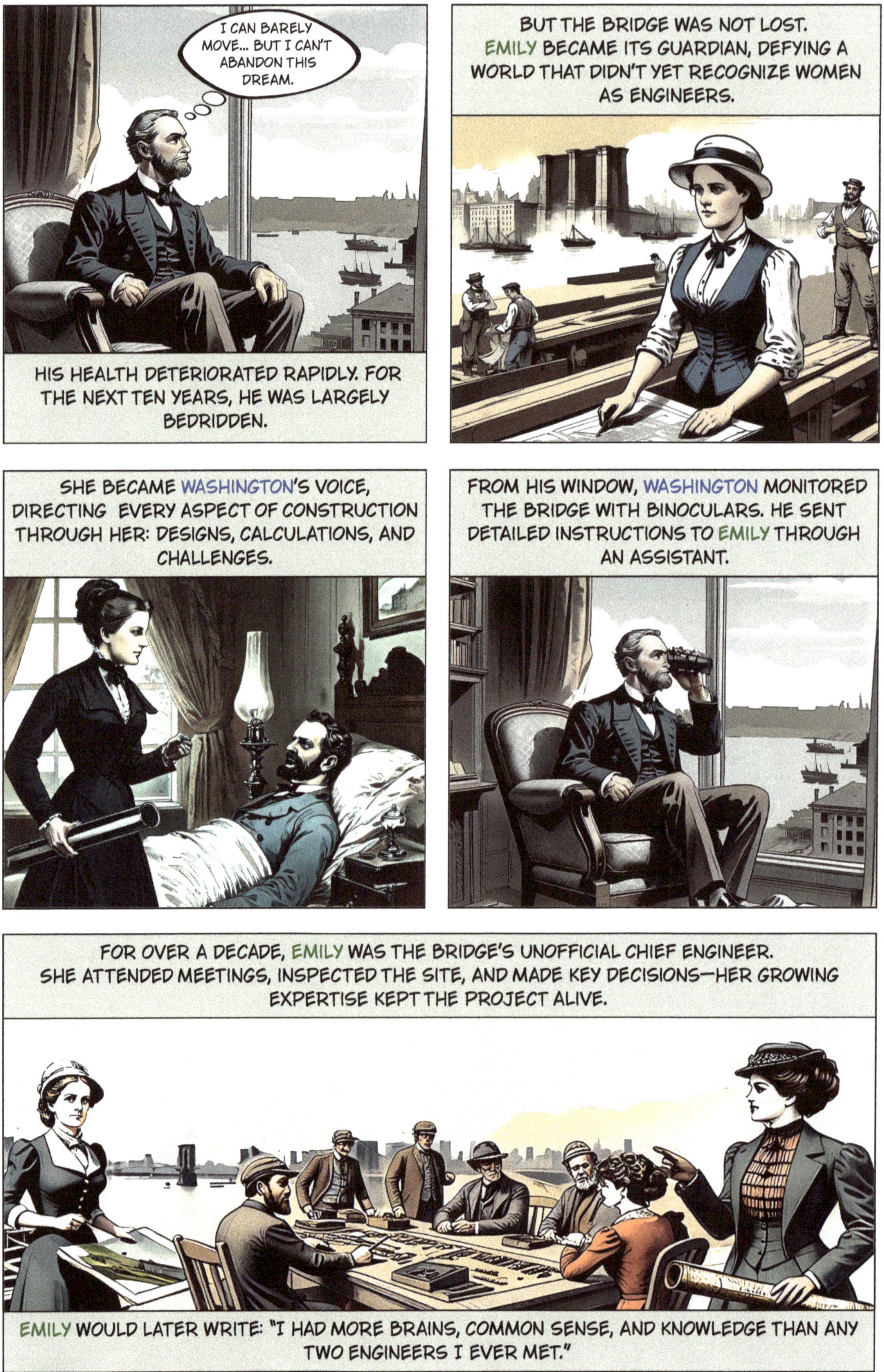
I CAN BARELY MOVE... BUT I CAN'T ABANDON THIS DREAM.
HIS HEALTH DETERIORATED RAPIDLY. FOR THE NEXT TEN YEARS, HE WAS LARGELY BEDRIDDEN.
BUT THE BRIDGE WAS NOT LOST. EMILY BECAME ITS GUARDIAN, DEFYING A WORLD THAT DIDN'T YET RECOGNIZE WOMEN AS ENGINEERS.
SHE BECAME WASHINGTON'S VOICE, DIRECTING EVERY ASPECT OF CONSTRUCTION THROUGH HER: DESIGNS, CALCULATIONS, AND CHALLENGES.
FROM HIS WINDOW, WASHINGTON MONITORED THE BRIDGE WITH BINOCULARS. HE SENT DETAILED INSTRUCTIONS TO EMILY THROUGH AN ASSISTANT.
FOR OVER A DECADE, EMILY WAS THE BRIDGE'S UNOFFICIAL CHIEF ENGINEER. SHE ATTENDED MEETINGS, INSPECTED THE SITE, AND MADE KEY DECISIONS—HER GROWING EXPERTISE KEPT THE PROJECT ALIVE.
EMILY WOULD LATER WRITE: "I HAD MORE BRAINS, COMMON SENSE, AND KNOWLEDGE THAN ANY TWO ENGINEERS I EVER MET."

IN MARCH 1875, ONE OF THE CAISSONS ACCIDENTALLY DEPRESSURIZED, TRIGGERING A MASSIVE EXPLOSION. WATER, MUD, AND STONE SHOT 152 METERS INTO THE AIR.
SEVERAL WORKERS DIED, MANY WERE INJURED. THE TRAGEDY WEIGHED HEAVILY ON THOSE IN CHARGE—BUT THE BRIDGE HAD TO GO ON.
BY THE TIME THE BRIDGE IS COMPLETE, 27 WORKERS WILL HAVE LOST THEIR LIVES.
BY 1876, THE MOST CHALLENGING PHASE OF THE CONSTRUCTION —THE TOWERS— WAS FINALLY COMPLETED.
THEN, WORK BEGAN ON THE STEEL CABLES!—MAKING THE BROOKLYN BRIDGE THE FIRST SUSPENSION BRIDGE TO USE STEEL CABLES—A MATERIAL THAT WOULD SOON CHANGE ENGINEERING WORLDWIDE.
EACH OF THE BRIDGE'S FOUR MASSIVE CABLES CONTAINED OVER 5,000 WIRES, TIGHTLY BUNDLED INTO STRANDS FOR MAXIMUM STRENGTH.

BUT BEHIND THE SCENES, CORRUPTION CAST A SHADOW OVER THE BRIDGE.
WIRE SUPPLIER J. LLOYD HAIGH CERTIFIED HIS CABLES AS TOP QUALITY—YET SECRETLY DELIVERED THOUSANDS OF METERS OF SUBSTANDARD WIRE, POCKETING THE PROFITS.
$
HEH, HEH!
TOP QUALITY WIRE
BY 1878, WHEN THE FRAUD WAS DISCOVERED, IT WAS TOO LATE TO REPLACE ALL THE FAULTY WIRES. SO, WASHINGTON HAD NO CHOICE BUT TO ORDER EXTRA CABLES TO REINFORCE THE BRIDGE.
BUT WHEN NEWSPAPERS EXPOSED THE SCANDAL, PUBLIC FEAR GREW.
WOULD IT BE ENOUGH?
COULD THE BRIDGE TRULY CARRY THE IMMENSE WEIGHT IT WAS DESIGNED FOR?
BRIDGE SAFETY IN QUESTION: FAULTY WIRES FOUND!
AFTER 14 YEARS OF PLANNING AND CONSTRUCTION—AND 11 YEARS OF EMILY'S LEADERSHIP—THE BROOKLYN BRIDGE WAS FINALLY COMPLETE.
ON MAY 17, 1883, IN A PRIVATE CEREMONY, EMILY AND HER SON CROSSED THE BRIDGE FOR THE FIRST TIME, CARRYING A WHITE ROOSTER —A SYMBOL OF VICTORY.

SEVEN DAYS LATER, AS THE BRIDGE OFFICIALLY OPENED, EMILY CROSSED IT AGAIN. THAT HISTORIC DAY SAW 150,000 PEOPLE —INCLUDING PRESIDENT CHESTER ARTHUR—AND 1,800 VEHICLES MAKING THE CROSSING.
CONGRESSMAN ABRAM STEVENS HEWITT HONORED EMILY, CALLING THE BRIDGE "AN IMPERISHABLE MONUMENT TO A WOMAN'S DEVOTION AND SACRIFICE...".
WE JUST CROSSED MORE THAN A BRIDGE.
WITH A MAIN SPAN OF 487 METERS, THE BROOKLYN BRIDGE WAS THE LONGEST SUSPENSION_BRIDGE EVER BUILT.
THAT NIGHT, FIREWORKS ILLUMINATED THE SKY AS THOUSANDS GATHERED ON BOTH SHORES TO CELEBRATE....
LOOK AT IT SHINE!
...WHILE WASHINGTON WATCHED FROM HIS WINDOW, TAKING IN EVERY DETAIL.

Based on a historical photograph from Internet Archive (by Tekniska museet; contributor: Arkivkopia).

SIX DAYS AFTER THE INAUGURATION, ON MAY 30, 1883, MEMORIAL DAY CROWDS FILLED THE BRIDGE.
SUDDENLY, A WOMAN STUMBLED—SOMEONE SCREAMED, "THE BRIDGE IS COLLAPSING!".
IT'S FALLING!
OH NO!
THE BRIDGE IS COLLAPSING!
PANIC SPREAD LIKE WILDFIRE. WITHIN SECONDS, PEOPLE RUSHED TO ESCAPE, CERTAIN THE BRIDGE WAS COLLAPSING.
RUN!
A WORKER TORE DOWN A SECTION OF THE IRON FENCE TO CREATE AN EXIT...
BUT IT WAS TOO LATE—12 PEOPLE DIED, AND OVER 35 WERE INJURED. YET, THE BRIDGE HAD NEVER BEEN IN DANGER.
DOUBTS ABOUT THE BRIDGE'S STRENGTH LINGERED FOR MONTHS.
?
?
?
IN 1884, TO SILENCE CRITICS, BARNUM WAS FINALLY ALLOWED TO PROVE HIS POINT.
OH JOY. CROSSING TIME.

LED BY THE FAMOUS ELEPHANT JUMBO, BARNUM'S CIRCUS ANIMALS MARCHED PROUDLY ACROSS THE BRIDGE.
TOGETHER, THE ANIMALS WEIGHED 5 TONS.
WITH EACH STEP, FEARS FADED—THE BRIDGE WAS BUILT TO LAST.

BARNUM'S STUNT GRABBED HEADLINES, CEMENTING THE BRIDGE'S REPUTATION AS AN ENGINEERING MARVEL.
Brooklyn Bridge
HOLDS 21 ELEPHANTS, 7 CAMELS, 10 DROMEDARIES..
WIRE FRAUDSTER HAIGH'S REPUTATION WAS RUINED—HE WOULD LATER SERVE TIME FOR OTHER FRAUDULENT CRIMES.

MORE THAN 140 YEARS LATER, THE BROOKLYN BRIDGE STILL STANDS—AN ENDURING SYMBOL OF AMBITION AND INNOVATION.

WHEN IT OPENED, IT WAS THE LONGEST SUSPENSION BRIDGE IN THE WORLD, AND THE FIRST BRIDGE TO USE STEEL CABLES.

SINCE THE 1980S, FLOODLIGHTS HAVE ILLUMINATED ITS NEO-GOTHIC TOWERS, MAKING THEM GLOW EACH NIGHT ABOVE THE EAST RIVER.

Main cables arch above, with vertical wires holding the bridge deck.

A LANDMARK OF ENGINEERING AND URBAN TRANSFORMATION

The Brooklyn Bridge set new engineering standards with steel cables and "parallel wire construction". It also pioneered the use of large pneumatic caissons, enabling safe underwater excavation on an unprecedented scale.

Beyond its engineering significance, the Brooklyn Bridge became a symbol of **progress** and **unity**. Built during the inequalities of the Gilded Age, it was seen as a collective achievement, representing ambition and the ability to overcome both physical and social barriers. The connection between Manhattan and Brooklyn stood as a testament to society's capacity to work together and bridge its differences.

Vertical and diagonal cables reveal its hybrid structure—part suspension, part cable-stayed.

Author's Note

I've walked across the Brooklyn Bridge many times in my life—at sunrise, in winter, in summer. Its sounds and presence never felt the same twice.

I first saw New York before ever visiting Buenos Aires, the capital of my own country. I came to study English, living in Brooklyn while attending classes in Manhattan. The bridge has stayed with me ever since. The photos around this text are ones I took during my last visit.

RECOGNITION AND LEGACY

In 1964, the American Society of Civil Engineers (ASCE) named it an **International Historic Civil Engineering Landmark**, honoring its groundbreaking design and its pivotal role in civil engineering history.

Since then, it has also been recognized as a **National Historic Landmark** in the United States.

In 2017, the bridge was added to the **National Register of Historic Places** as part of the "New York City Historic Bridges" group, further recognizing its significance in the city's architectural and engineering legacy.

Iconic neo-Gothic towers anchoring the steel-wire suspension of the bridge. Strength and elegance.

Once known as the New York and Brooklyn Bridge, and later as the East River Bridge, it was officially renamed the Brooklyn Bridge in 1915.

"A bridge is a poem stretched across a river, a symphony of stone and steel."

– Engineer D. B. Steinman, "Brooklyn Bridge – Nightfall" (in the 1950s).

The current pathways for both vehicles and pedestrians were remodeled by the engineer and bridge builder, D.B. Steinman, in 1948, to increase road capacity.

Sources & References

For this documentary comic, I relied on invaluable technical details regarding cable setting and the caisson method, sourced from the Catskill Archive website: (https://catskillarchive.com/rrextra/bbstory.html).

Insights into the Roebling family were drawn from the Archives and Special Collections at Rutgers University (https://archives.libraries.rutgers.edu/repositories/11/resources/908).

Additionally, I consulted the blog Today in History – Mr. Roebling's Bridge, which provided historical perspective on the bridge's early vision and approval (https://todayinhistory.blog/2018/01/03/january-3-1870-mr-roeblings-bridge/).

The VinePair article "Why 21 Elephants and 17 Camels Once Crossed the Brooklyn Bridge" (https://vinepair.com/cocktail-chatter/brooklyn-bridge-greatest-showman-animals/), provided key details for one of the final scenes.

Historical photographs were sourced from Internet Archive, and Wikipedia.

David McCullough's book provides an extensively researched account of the bridge's history, covering the challenges, innovations, and scandals, such as the J. Lloyd Haigh wire fraud incident.

McCULLOUGH, D. (1972). "The Great Bridge: The Epic Story of the Building of the Brooklyn Bridge". Simon & Schuster, New York, 636 pp.

The Brooklyn Bridge was more than a marvel of engineering—it became a symbol of perseverance, sacrifice, and human ingenuity. When it opened, it united two great cities, silenced the skeptics, and redefined what bridges—and people—could do.

Decades later, across the ocean, a new project would rise—not in the heart of a city, but high in the unforgiving Alps. There, engineers carved concrete into the depths of the mountains, building one of the tallest dams ever conceived.

The mountains watched in relative silence.

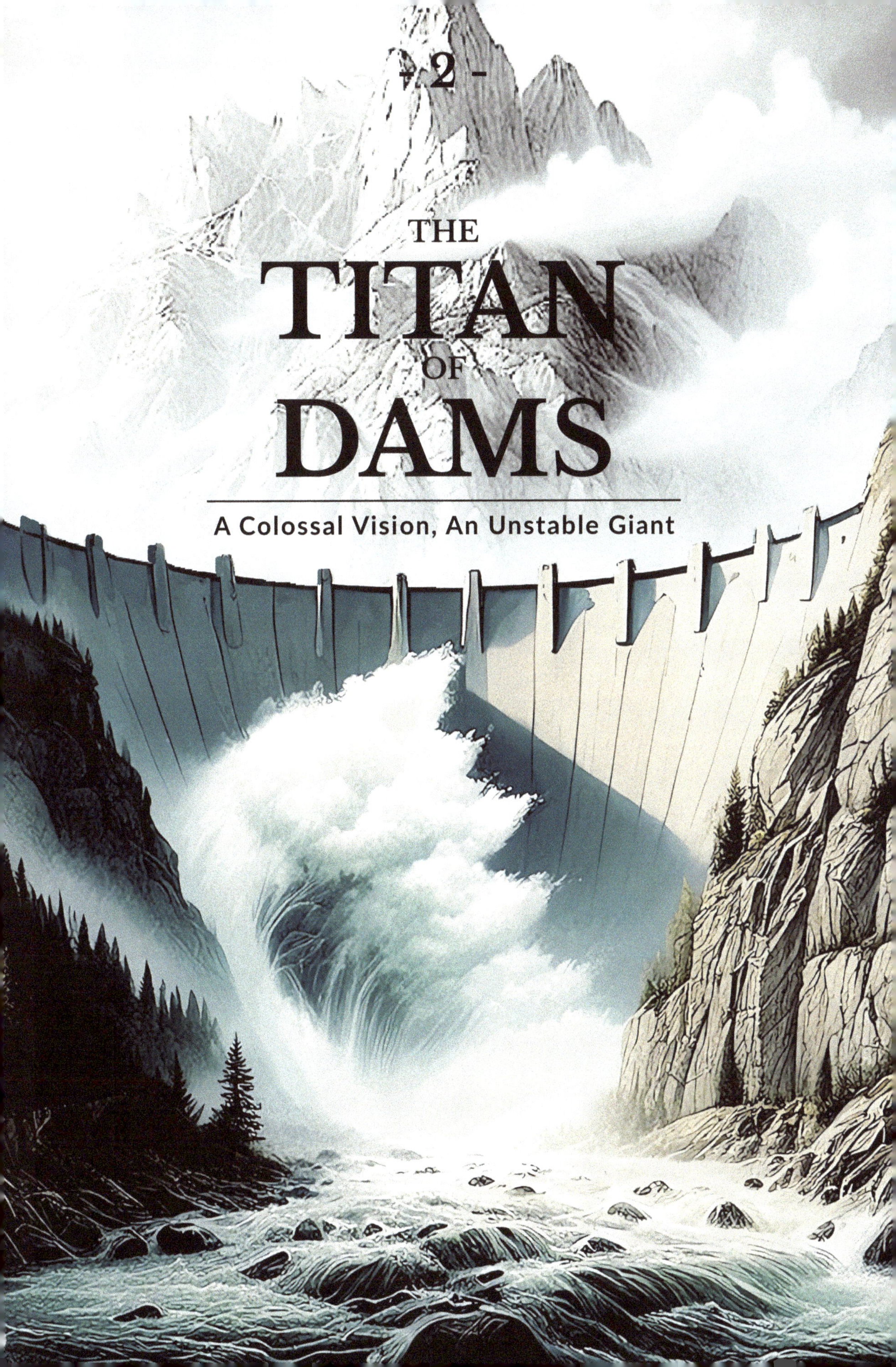
- 2 -
THE
TITAN
OF
DAMS
A Colossal Vision, An Unstable Giant

CARLO SEMENZA WAS BORN IN 1893 IN MILAN, ITALY.
HE GREW UP WITH TWO LIFELONG PASSIONS: MOUNTAINEERING AND HYDRAULIC ENGINEERING.
IN 1929, HE MADE HIS FIRST TRIP TO VAJONT VALLEY, SEARCHING FOR THE PERFECT SPOT TO BUILD A DAM.
"VAJONT" MEANS "GOING DOWN" IN LADIN.
A STEEP CANYON IN THE DOLOMITES OF THE ITALIAN ALPS SEEMED LIKE THE IDEAL SPOT.
ROCKS ARE VERY GOOD... LIMESTONE IS HONEST—IT SHOWS ITS FLAWS RIGHT ON THE SURFACE.
BY THEN, A PRIVATE COMPANY, THE ADRIATIC ELECTRICITY COMPANY (S.A.D.E.), HAD HIRED SEMENZA TO DESIGN THE DAM.
SADE
VENEZIA
THE COMPANY BELONGED TO GIUSEPPE VOLPI, A POWERFUL BUSINESSMAN AND FOUNDER OF THE VENICE FILM FESTIVAL, ALSO KNOWN AS THE COUNT OF MISURATA. VOLPI ENTRUSTED SEMENZA TO BRING THE AMBITIOUS "GRAN VAJONT PROJECT" TO LIFE.
VAJONT VALEY DAMS
IN 1943, VOLPI CELEBRATED THE PROJECT'S APPROVAL, WHICH INCLUDED FOUR DAMS AS PART OF A VAST HYDROELECTRIC NETWORK.
THIS AMBITIOUS PROJECT PROMISED TO BRING POWER TO COUNTLESS HOMES ACROSS NORTHEASTERN ITALY.
VOLPI PASSED AWAY IN 1947, BUT THE PROJECT WAS ALREADY MOVING FORWARD.

MEANWHILE, LOCALS VOICED CONCERNS ABOUT THE UNSTABLE MONTE TOC, WHERE ONE OF THE FOUR DAMS, THE VAJONT DAM, WOULD BE BUILT.
THEY EVEN CALLED IT "THE MOUNTAIN THAT WALKS" DUE TO THE FREQUENT LANDSLIDES.
LOCAL RESIDENTS PROTESTED FIERCELY, WARNING THAT DISASTER WAS IMMINENT.
NO DAM ON MONTE TOC!
NO DAM ON
DESPITE THE MOUNTING WARNINGS, CONSTRUCTION WENT AHEAD ANYWAY IN 1957...
...ALONG WITH FORCED LAND EXPROPRIATIONS. S.A.D.E. PRESSURED LOCALS TO QUICKLY SURRENDER THEIR LAND.
HEH, HEH!
AT THAT TIME, SEMENZA WAS ONE OF THE WORLD'S TOP DAM DESIGNERS.
Longarone
Casso
Erto
VAJONT LAKE
FUTURE DAM
A CURVED DOUBLE-ARCHED DAM!
IN THE FINAL DESIGN REVISION, S.A.D.E. RAISED THE DAM'S HEIGHT BY 60 METERS—MAKING IT THE TALLEST DAM IN THE WORLD AT 262 METERS!

THEN, IN MARCH 1959, A MASSIVE LANDSLIDE DUMPED 3 MILLION CUBIC METERS OF ROCK INTO NEARBY PONTESEI LAKE.
ONE MAN DIED.
THIS RAISED FEARS OF CATASTROPHE, AND THE PRESS AMPLIFIED LOCAL CONCERNS.
JOURNALIST TINA MERLIN, FROM THE NEWSPAPER L'UNITA, BECAME THEIR VOICE, DEFYING PRESSURE FROM AUTHORITIES AND INDUSTRY.
SEMENZA KNEW A DEEPER GEOLOGICAL INVESTIGATION WAS NECESSARY.
?
HE CALLED IN AUSTRIAN GEOMECHANICS EXPERT PROFESSOR LEOPOLD MÜLLER TO ASSESS MONTE TOC.
IN JULY 1959, PROF. MÜLLER VISITED THE DAM SITE AND ENLISTED EDOARDO SEMENZA, CARLO'S SON, TO INVESTIGATE FURTHER.
EDOARDO WAS A RISING STAR IN GEOLOGY.
HE SOON IDENTIFIED EVIDENCE OF A PREHISTORIC LANDSLIDE ON MONTE TOC—LIKELY TO REACTIVATE WITH THE DAM.
I SHOULD TELL PROFESSOR MÜLLER

IN OCTOBER 1959, EDOARDO AND HIS FATHER TRAVELED TO AUSTRIA TO MEET PROF. MÜLLER.
SLOWLY
MÜLLER BELIEVED THE LANDSLIDE WOULD UNFOLD GRADUALLY... SLOW LIKE A GLACIER.
BUT EDOARDO WAS CONVINCED THE WHOLE MOUNTAIN BLOCK MIGHT SHIFT ALL AT ONCE.
IN BLOCK
WHILE SOME EXPERTS DISMISSED EDOARDO, MONTE TOC WAS PUT UNDER CLOSE MONITORING.
WELLS AND PIEZOMETERS WERE INSTALLED TO START TRACKING SLOPE MOVEMENTS.
BY SEPTEMBER 1960, THE WORLD'S TALLEST DAM HAD BEEN COMPLETED, AND THE RESERVOIR BEGAN FILLING.
ENGINEER MARIO PACINI WAS TASKED WITH MONITORING THE DAM AND LANDSLIDE RISKS.
A RESPECTED ENGINEER FROM VENICE, PACINI HAD WORKED WITH SEMENZA BEFORE.

FORTY DAYS LATER, AS THE WATER LEVEL HIT 645 METERS, A 700,000-CUBIC-METER LANDSLIDE CRASHED INTO THE LAKE.
IT WAS JUST A FRACTION OF THE ANCIENT LANDSLIDE.
OPERATORS QUICKLY LOWERED THE WATER LEVEL, AND FOR A TIME, MOVEMENT CEASED.
BUT A NEW 2 KM CRACK HAD APPEARED ON MONTE TOC, AND PACINI HAD TO KEEP A CLOSE EYE ON IT.
WEEKS LATER, JOURNALIST TINA MERLIN WAS ACQUITTED OF "SPREADING ALARM" ABOUT THE DAM. SHE HAD BEEN RIGHT ALL ALONG.
GIVEN THE CLEAR LANDSLIDE THREAT, SEMENZA AND PACINI DESIGNED A 2-KILOMETER (5-METER-WIDE) BYPASS TUNNEL.
THE TUNNEL WOULD LET WATER FLOW FROM THE UPPER TO LOWER LAKE IF ANOTHER LANDSLIDE HIT.

SEMENZA HAD ONE MORE STRATEGY: HE COMMISSIONED A SCALE MODEL OF THE DAM FROM THE UNIVERSITY OF PADUA.
1:200 SCALE
A BIG DEAL AT THE TIME!
ONCE THE MODEL WAS READY, IT SHOWED THE DAM COULD WITHSTAND THE EXPECTED LANDSLIDE—AS LONG AS THE LAKE'S WATER LEVEL STAYED BELOW 700 METERS.
IT WAS ITALY'S FIRST PHYSICAL MODEL AND ONE OF THE EARLIEST WORLDWIDE.
IN FEBRUARY 1961, PROF. MÜLLER PRESENTED HIS 15TH REPORT, WARNING ABOUT THE LACK OF MEASURES TO CONTROL POTENTIAL LANDSLIDES FROM MONTE TOC.
DAM
MULLER
BY APRIL 1961, SEMENZA, UNDER MOUNTING PRESSURE AND FULLY AWARE OF THE RISKS, WAS DEEPLY TROUBLED.
"AFTER SO MANY SUCCESSFUL PROJECTS... I NOW FACE SOMETHING SO VAST IT SEEMS BEYOND MY CONTROL".
IN OCTOBER 1961, AS A CRUCIAL DECISION LOOMED AND HIS SON QUESTIONED THE EXPERTS, SEMENZA SUFFERED A FATAL STROKE.
HIS SUCCESSOR, BIADENE, WAS DETERMINED TO COMPLETE FILLING THE LAKE—HE ABANDONED THE MODEL'S FINDINGS AND DISMISSED PROF. MÜLLER'S REPORT.
DAM

IN NOVEMBER 1962, THE LAKE REACHED 700 METERS, AND THE DAM WAS OFFICIALLY INAUGURATED.
BY MARCH 1963, THE ITALIAN GOVERNMENT NATIONALIZED S.A.D.E. UNDER ENEL (ITALY'S NATIONAL ELECTRICITY AGENCY).
I'LL RAISE THE LEVEL TO 715 M.
THE DAM'S FINAL PRICE DEPENDED ON HOW FULL THE LAKE WAS—715 METERS MEANT MAXIMUM VALUE!
BY SEPTEMBER 1963, AS WATER LEVELS REACHED 710 METERS, LANDSLIDE RISK INTENSIFIED—OPERATORS RUSHED TO LOWER IT BACK TO 700.
CREAK!
CREAK!
CREAK!
CREAK!
YET THIS TIME, THE GROUND KEPT MOVING : 1 CM/DAY ON SEPTEMBER 18, 15 CM/DAY BY SEPTEMBER 25. 30 CM/DAY BY OCTOBER 6, 40 CM/DAY BY OCTOBER 8.
PANIC SET IN! ON THE MORNING OF OCTOBER 9: BIADENE SENT AN URGENT LETTER, PLEADING FOR PACINI TO RETURN FROM AMERICA.
Water Level (meters)
700
600
by-pass build
710m
Moving (cm/day)
3
0
"P.D.: MAY GOD SEND US GOOD LUCK".
THAT NIGHT, MILLIONS WATCHED THE EUROPEAN CHAMPIONS CUP MATCH ON TV.
5
REAL MADRID VS. GLASGOW RANGERS.

JUST A FEW KILOMETERS AWAY, AT 10:39 PM, MONTE TOC PLUNGED INTO VAJONT LAKE AT 30 METERS PER SECOND (108 KM/H).
NO WARNING.

THE LANDSLIDE UNLEASHED A COLOSSAL WAVE—MORE THAN 200 METERS HIGH—THAT SURGED OVER THE DAM
FIFTY MILLION CUBIC METERS OF WATER WERE PROPELLED FORWARD—ENOUGH TO FILL 20,000 OLYMPIC-SIZE SWIMMING POOLS.

THE TOWERING WAVE ENGULFED LONGARONE, ERASING 80% OF THE TOWN AND DEVASTATING NEARBY VILLAGES.
HAILED AS THE "TITANIC OF DAMS", VAJONT CLAIMED OVER 2,000 LIVES—WITHOUT EVER GENERATING A SINGLE WATT OF ELECTRICITY.
A PORTION OF MONTE TOC WAS ULTIMATELY BURIED IN THE LAKE.
AND YET, THE DAM REMAINED INTACT—A SILENT WITNESS TO THE TRAGEDY.
PACINI, TORMENTED BY GUILT FOR THE DISASTER AND THE PEOPLE HE HAD KNOWN...
"GUILT OR NOT, TWO THOUSAND ARE DEAD."
THE NIGHT BEFORE THE 1968 TRIAL, PACINI'S LAWYERS WAITED FOR HIM AT VENICE'S SANTA LUCIA STATION.

VAJONT'S LANDSCAPE TODAY

The Vajont Dam:
Still standing tall, the Vajont Dam remains a stark reminder of the tragedy. Though no longer in use, it endures as a historical monument.

Monte Toc:
The colossal landslide on Mount Toc left a lasting scar—nature's fury etched into the landscape. It stands as a sobering testament to human mismanagement. Today, the gorge lies still, untouched—its silence echoing the tragedy.

ECHOES OF DISASTER

Longarone:
Nearly obliterated by the wave, Longarone has since been rebuilt into a modern town. Yet the memory of the tragedy remains deeply ingrained. Memorials, museums, and annual commemorations honor the victims and preserve their story.

Annual Commemorations:
Every October 9, the Vajont community gathers to commemorate the disaster. Ceremonies include candlelit vigils, educational events, and activities to honor the victims and raise awareness.

Sources & References

Throughout the creation of this project, I came across various sources that shed light on crucial aspects, from the early sketches to the dramatic events that unfolded. Among the most insightful were the works of Müller-Salzburg (1987) and Barrotta & Montuschi (2018), which offered a unique visual perspective on the Vajont landslide mechanics, the bypass tunnel design, and the scale model developed at the University of Padua.

Another valuable source was the article by O. Domosti in Jot Down Cultural Magazine (https://jotdown.es/2012/06/vajont-el-titanic-de-las-presas/), the paper by Ward & Day (2011) and of course, Wikipedia.

It's important to note that the 1960 report prepared for SADE by Edoardo Semenza—predicting the possibility of a landslide—was co-authored with geologist Francesco Guidici, who is not featured in this documentary comic.

MULLER-SALZBURG, L. (1987). "The Vajont catastrophe - a personal review." Engineering Geology, 24, pp. 423–444. Elsevier Science Publishers B.V., Amsterdam.
BAROTTA, P. & MONTUSCHI, E. (2018). "Chapter 1. The dam project: Who are the experts?: A philosophical lesson from the Vajont disaster". Science and Democracy: Controversies and conflicts, edited by Pierluigi Barrotta & Giovanni Scarafile. John Benjamins Publishing Company, pp. 17–34.
WARD, S. & DAY, S. (2011). "The 1963 Landslide and Flood at Vaiont Reservoir Italy. A tsunami ball simulation". Italian Jorunal of Geosciences, 130, pp. 16-26.

The Vajont disaster was a stark reminder that even advanced engineering projects can be undone by nature. A tragedy where ambition collided with oversight, leaving behind a silent monument to everything that went wrong.

Decades earlier, and an ocean away, another monumental structure was taking shape—not in the Alps, but in the rugged sierras of Cordoba, Argentina. By regulating the flow of the mountain rivers, an ambitious dam would not only tame the waters, but also unlock the region's potential for growth and prosperity, fueling the economic rise of central Argentina.

But engineering was only part of the challenge. Political struggles, clashing interests, and deep-rooted skepticism cast long shadows over its construction.

Would those who dared to push the limits—and build the world's largest reservoir—pay the highest price?

- 3 -

THE COLOSSUS OF DOUBTS

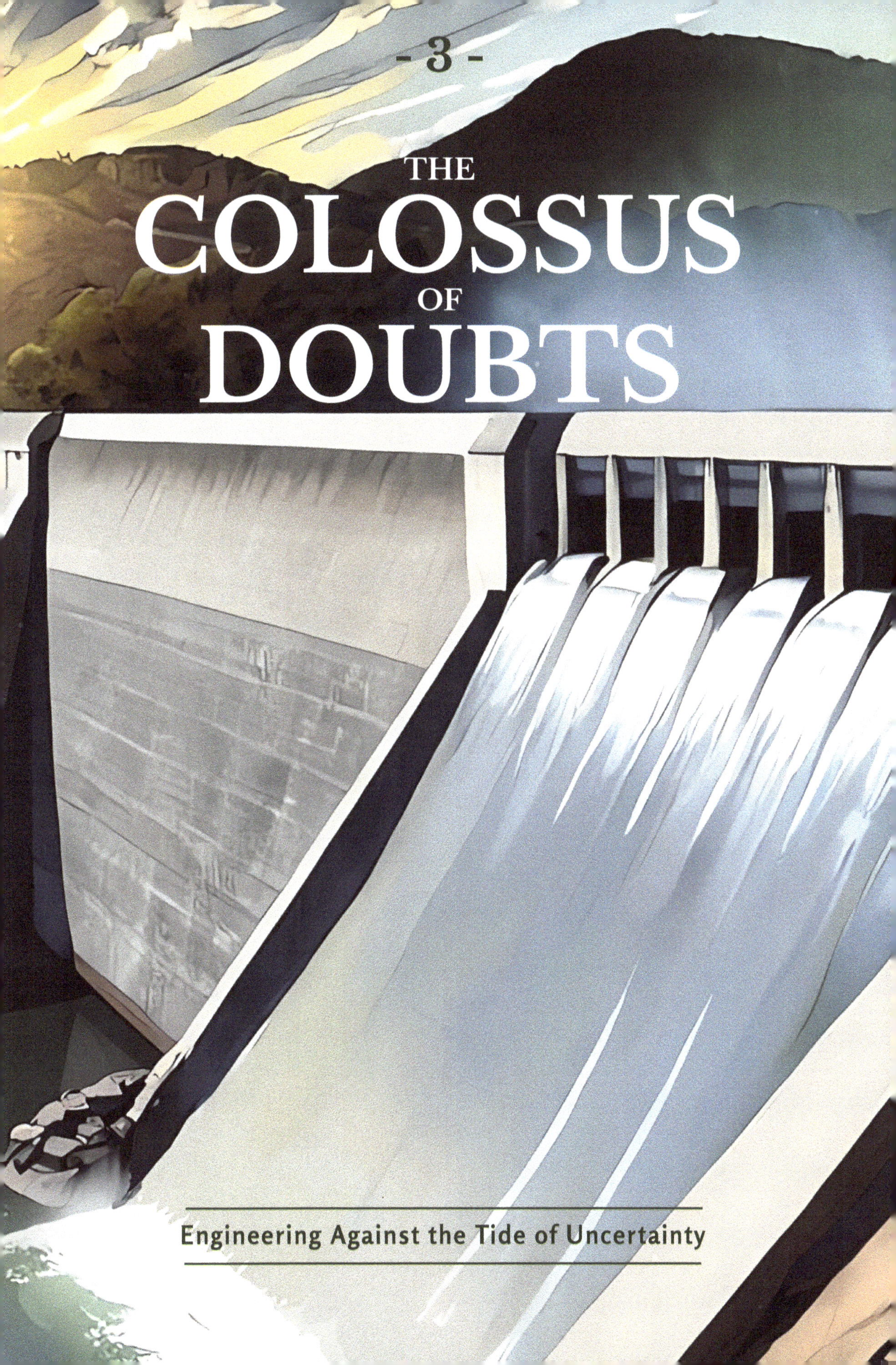

Engineering Against the Tide of Uncertainty

BENEATH THESE WATERS LIES SAN ROQUE CHAPEL, THE BIRTHPLACE OF ARGENTINA'S FIRST BLESSED, MARIA CABANILLAS.
THIS WAS NO ORDINARY LAKE.
THE WORLD HAD NEVER SEEN ANYTHING LIKE IT. AT THE TIME, IT WAS THE LARGEST MAN-MADE LAKE EVER BUILT AND SOUTH AMERICA'S FIRST DAM. ON SEPTEMBER 8, 1891, THE DAM OFFICIALLY OPENED.
LOCALS CALLED IT "THE COLOSSUS".

BUT WHAT ABOUT THE MAN WHO STOOD PROUDLY ATOP THE DAM, EVEN AS DOUBTS AND TENSIONS ROSE?

HIS NAME WAS CARLOS ADOLFO CASAFFOUSTH, BORN IN BUENOS AIRES, 1854, INTO A FAMILY WHERE BOOKS AND IDEALS SHAPED DESTINIES.
HIS GODFATHER? DOMINGO FAUSTINO SARMIENTO—ARGENTINA'S REVOLUTIONARY PRESIDENT (1868-1874) AND ARCHITECT OF THE EDUCATION SYSTEM.
FROM A YOUNG AGE, CASAFFOUSTH SHOWED A PASSION FOR MATHEMATICS, AND IN 1888, HE WAS SENT TO THE CENTRAL SCHOOL IN PARIS...
...WHERE HE EARNED HIS CIVIL ENGINEERING DEGREE AND CROSSED PATHS WITH FRENCH ENGINEER GUSTAVE EIFFEL.
WITH HIS DEGREE IN HAND, CASAFFOUSTH RETURNED TO ARGENTINA, EAGER TO TAKE ON MAJOR PROJECTS.
TRANS-ANDEAN RAILWAY (1880-1883).
HIS MISSION? TO IMPROVE ARGENTINA'S CONNECTIVITY.
BUT IN 1881, CASAFFOUSTH ARRIVED IN CORDOBA, IN THE HEART OF THE COUNTRY, AND WAS CAPTIVATED BY ITS POTENTIAL.
THIS CITY IS EXPANDING SO RAPIDLY...

WITH A BOOMING POPULATION AND EXPANDING AGRICULTURE, CORDOBA NEEDED A STEADY WATER SUPPLY.
Cordoba
RAIN IS A SUMMER FRUIT!
IT WAS IN 1886 THAT CASAFFOUSTH EMBARKED ON AN AMBITIOUS PROJECT—A DAM TO SECURE WATER FOR CORDOBA'S 50,000 RESIDENTS, IRRIGATE SURROUNDING FARMLAND, AND HELP REGULATE THE CITY'S DESTRUCTIVE FLOODS.
WITH THIS, WE'RE ENSURING CORDOBA'S FUTURE.
THE BIG CONTRACT
CASAFFOUSTH SOON JOINED FORCES WITH BIALET MASSE—A MAN OF MANY TALENTS, READY TO HELP TRANSFORM CORDOBA'S FUTURE.
B. MASSE WAS A SPANISH DOCTOR AND LAWYER, WHO HAD ARRIVED IN CORDOBA IN 1877, BACKED BY POWERFUL CONNECTIONS.
MARRIED TO ZULEMA LAPRIDA, GRANDDAUGHTER OF A SIGNER OF ARGENTINA'S INDEPENDENCE, HE WAS DEEPLY TIED TO THE COUNTRY'S ELITE.
WHILE STUDYING THE LIME ON HIS LAND, B. MASSE DISCOVERED IT WAS NATURALLY "HYDRAULIC"—A RARE FIND THAT HARDENED UPON CONTACT WITH WATER.
!!
A DREAM MATERIAL FOR CONSTRUCTION!

EVEN THE JESUITS KNEW ITS VALUE. CENTURIES EARLIER (1599–1767), THEY USED THIS WORLD-CLASS LIME TO BUILD IN CORDOBA—AND MUCH OF IT STILL STANDS TODAY.
BY 1886, B. MASSE'S "LA PRIMERA ARGENTINA" FACTORY—THE COUNTRY'S FIRST INDUSTRIAL LIME PRODUCTION SITE—WAS OPERATING AT FULL SPEED.
THE LIME WAS FIRED IN KILNS LINED WITH "PIEDRA SAPO", A LOCAL STONE THAT COULD WITHSTAND TEMPERATURES OVER 1100°C, ESSENTIAL FOR PRODUCING HIGH-QUALITY LIME FOR THE DAM.
UNTIL THEN, ARGENTINA HAD RELIED ON IMPORTED MATERIALS—FRENCH LIME, ENGLISH CEMENT... NOT ANYMORE!
ONCE FIRED, THE LIME TRAVELED 86 METERS THROUGH TUNNELS ON MINING CARTS.
USING CUTTING-EDGE TECHNOLOGY FOR HIS TIME, B. MASSE PRODUCED TOP-QUALITY PRODUCTS—RIGHT IN THE HEART OF CORDOBA.
THE DAM PROJECT BEGAN—BUT FIRST, LAND HAD TO BE TAKEN, INCLUDING THE CEMETERY AND THE SAN ROQUE CHAPEL, NAMED AFTER A SAINT OFTEN DEPICTED WITH A LOYAL DOG AT HIS SIDE.
SAN ROQUE

THE DAM PROJECT WAS PUT TO THE TEST: LAND DISPUTES, POLITICAL OPPOSITION, MEDIA ATTACKS—EVEN A CHOLERA OUTBREAK.
YET, AFTER 3 YEARS AND THE RELENTLESS EFFORT OF 3,000 WORKERS—AND TONS OF LOCALLY PRODUCED LIME—THE SAN ROQUE DAM FINALLY STOOD: 115 M LONG AND 37 M HIGH.
THEN, THE POLITICAL LANDSCAPE SHIFTED—THE 1890 REVOLUTION FORCED PRESIDENT JUAREZ CELMAN, THE DAM'S KEY SUPPORTER, TO RESIGN.
DEMO-CRACY!
FREEDOM!
END ONE-MAN RULE!
WITH J. CELMAN GONE, THE DAM BECAME THE TARGET.
IT SHOULD HAVE BEEN BUILT WITH IMPORTED CEMENT!
CRITICS, DRIVEN BY POLITICS AND PROFIT, CONDEMNED THE DAM: "ONLY IMPORTED CEMENT SHOULD HAVE BEEN TRUSTED."
DOUBTS ABOUT THE DAM'S STABILITY GREW, AND FEAR BEGAN TO SPREAD. WATER LEVELS WERE KEPT LOW—FOR SAFETY—LEAVING THE DAM'S FUTURE UNCERTAIN.
?
IRRIGATION WAS HALTED, AND WATER ONLY SUPPLIED CORDOBA CITY, LEAVING FARMS DRY. TENSIONS SOARED.

YET, ON SEPTEMBER 8, 1891, SOUTH AMERICA'S FIRST DAM—AND THE WORLD'S LARGEST MAN-MADE LAKE—WAS OFFICIALLY INAUGURATED, THOUGH WITH LOW WATER LEVELS AS A PRECAUTION.
JUST FOUR DAYS BEFORE THE OPENING, A RAILWAY WAS FINISHED, SPARKING A SURGE IN TOURISM.
THE CENTRAL CEREMONY TOOK PLACE ON THE DAM WALL, IN FRONT OF A BUST OF FORMER PRESIDENT J. CELMAN, THE DAM'S KEY SUPPORTER, WHO HAD RESIGNED A YEAR EARLIER.

FRESH FROM UNVEILING HIS TOWER IN PARIS, THE FAMED ENGINEER GUSTAVE EIFFEL REPORTEDLY SAID: "TWO CONSTRUCTIONS ARE DRAWING THE WORLD'S ATTENTION RIGHT NOW: MY TOWER AND CORDOBA'S DAM. THE DIFFERENCE? THE DAM SERVES A PURPOSE!".

BUT DOUBTS ABOUT THE DAM PERSISTED IN ARGENTINA. IN JUNE 1892, THE NEW GOVERNMENT HIRED SWEDISH ENGINEER HJALMAR FREDERIK STAVELIUS TO ASSESS ITS SAFETY.
STAVELIUS, A PROMINENT ENGINEER IN ARGENTINA'S NATIONAL ENGINEERING OFFICE, HAD WORKED ON MAJOR PROJECTS, INCLUDING BUENOS AIRES' ICONIC "HOTEL DE INMIGRANTES".

ON JULY 27, 1892, AT 5:00 AM, A WAVE OF PANIC RIPPED THROUGH CORDOBA. POLICE CHARGED THROUGH THE STREETS, SHOUTING WARNINGS.
IT'S COMING DOWN!
THE DAM IS COLLAPSING!
PANICKED FAMILIES RUSHED FROM THEIR HOMES, DESPERATE TO ESCAPE THE CATASTROPHE THEY BELIEVED WAS INEVITABLE.

B. MASSE STOOD FROZEN OUTSIDE HIS HOME ON "CALLE ANCHA", THE WIDE ROAD, WATCHING IN ANGUISH AS THE CITY PLUNGED INTO CHAOS.
BUT... THE LAKE IS ALMOST DRY!
WHO COULD WANT THIS INFAMY, THIS HARM?

BY AFTERNOON, OFFICIALS CALLED IT A "FALSE ALARM"—BUT THE DAMAGE WAS DONE. TRUST IN THE DAM WAS SHATTERED.
El Porvenir
BOLETIN ESPECIAL

A MOTNH LATER, IN AUGUST 1892, THE GOVERNMENT UNVEILED THE HIGHLY ANTICIPATED STAVELIUS REPORT. A MASSIVE ENGINEERING DIAGRAM REVEALED MORE THAN 90 LEAKS—AND A DEEP CRACK RUNNING THROUGH THE DAM'S CORE.

STAVELIUS CONCLUDED THAT THE DAM—THE LARGEST OF ITS TIME, WITH A CAPACITY OF 260 MILLION CUBIC METERS—POSED AN IMMINENT RISK TO CORDOBA.

HIS SOLUTION? BUILD A SECONDARY WALL UPSTREAM, USING "ENGLISH MIX" CEMENT, AND FILLING THE GAP WITH THE SAME MATERIAL.

B. MASSE, A LAWYER HIMSELF, REFUSED TO BACK DOWN—HE FORMALLY REJECTED ANY LIABILITY FOR ALTERATIONS TO HIS STRUCTURE.

BUT THE GOVERNOR AT THAT TIME, PIZARRO, HAD ALREADY MADE UP HIS MIND: CASAFFOUSTH AND B. MASSE MUST BE ARRESTED!

BY OCTOBER 1892, THE TWO MEN WERE BEHIND BARS FOR "ENDANGERING LIVES AND WASTING PUBLIC FUNDS".

JAIL SHATTERED THEIR REPUTATIONS—AND THEIR FINANCES—BUT NEITHER GAVE UP.

FROM HIS PRISON CELL, B. MASSE WROTE A LETTER TO FORMER PRESIDENT J. CELMAN.

"I'M WRITING TO YOU FROM THE POLICE DEPARTMENT WHERE I AM IN PRISON FOR THE CRIME OF BUILDING THE DAM. I GIVE YOU MY WORD: THE DAM IS SOLID. IT WILL LAST—SO DON'T BE FRIGHTENED. YET THEY SAY IT MUST BE DEMOLISHED... SIMPLY BECAUSE IT WAS BUILT UNDER YOUR GOVERNMENT..."

DURING THE TRIAL, STAVELIUS'S QUALIFICATIONS CAME UNDER SCRUTINY—HIS ENGINEERING DEGREE WAS NEVER VALIDATED.
EVEN WORSE? EXPERTS FROM THE UNIVERSITY OF CORDOBA INSPECTED THE DAM. NO CRACKS WERE FOUND WHERE STAVELIUS HAD CLAIMED.

IN OCTOBER 1893, CASAFFOUSTH AND B. MASSE WERE FREED. THE NEXT DAY, GOVERNOR PIZARRO RESIGNED.
BY 1895, ALL CHARGES WERE DROPPED—BUT THEIR REPUTATIONS WERE DESTROYED.

BROKEN IN SPIRIT, CASAFFOUSTH LEFT CORDOBA AND RETREATED TO HIS WIFE'S ESTANCIA.
HE WOULD DIE IN 1900 AT JUST 46. YET, THE URBAN LEGEND OF HIS SUICIDE IN JAIL STILL LINGERS, STRONGLY, IN THE CORDOBA REGION.

DOUBTS ABOUT THE DAM'S STABILITY WOULDN'T GO AWAY WITH TIME. EVERY FLOOD REKINDLED FEAR... YET IT NEVER FAILED.

Based on a historical photograh from the Historical Archive of Cordoba.

THE NEW DAM HELD THE SAME AMOUNT OF WATER BUT INTRODUCED A GAME-CHANGER: A "MORNING GLORY" SPILLWAY—A FUNNEL-SHAPED DESIGN, LIKE AN "UPSIDE-DOWN FLOWER", TO SAFELY CHANNEL SUDDEN MOUNTAIN FLOODS.

WHAT ABOUT "THE COLOSSUS"?

AGAINST ALL ODDS... THE OLD DAM STILL STANDS!. NEITHER STORMS, NOR TIME, NOR EVEN HUMAN HANDS COULD BRING IT DOWN. EXPLOSIVES WERE USED IN AN ATTEMPT TO DESTROY IT—YET IT RESISTED. ONLY A SINGLE PASSAGE WAS EVER CARVED OPEN, ALLOWING WATER TO FLOW THROUGH.

A MONUMENT TO DEFIANCE, TO VISION, AND TO THE ENDURANCE OF THOSE WHO BUILT IT.

TODAY, THE NEW SAN ROQUE DAM CONTINUES TO SUPPLY CORDOBA, ARGENTINA'S SECOND-LARGEST CITY, WHILE ALSO HELPING TO CONTROL FLOODS.
MEANWHILE, THE OLD DAM'S WALL REMAINS VISIBLE MUCH OF THE YEAR.

The old dam's wall in 2025.

THE OLD SAN ROQUE DAM TODAY

In 2012, the old San Roque Dam ("The Colossus") was officially declared a **Heritage Site** by the Government of Cordoba, Argentina, recognizing its significance both as a functional engineering marvel and as a historical monument that reflects the region's development and the challenges of large-scale infrastructure projects.

THE KILN OF DR. BIALET MASSE

In 2018, the lime kiln built by Dr. Bialet Masse was declared a **National Historic Monument** and **Provincial Heritage Site**. Established in 1884 as part of Bialet Masse's factory, it continued to operate even after the bank seized Bialet Masse's assets in 1892, while he was in prison. In 1911, his sons purchased back the land, founding the **city of Bialet Masse**. Today, the kiln is a year-round must-see.

Kiln built by Dr. Bialet Masse in Cordoba.

Morning Glory spillway, a feature of the new San Roque Dam.

Current view of San Roque Reservoir.

Sources & References

In this documentary comic, an invaluable source for chronicling and providing context to the events was Luis Rodolfo Frias' book (1985). Equally useful were excerpts from Norberto Huber's book (2001), as well as websites like "La historia del embalse más grande del mundo" (https://diquesdecordoba.com.ar/san-roque-la-historia-del-embalse-mas-grande-del-mundo/), "San Roque Dam: A legend not so legend" (https://serargentino.com/en/people/the-worst-thing-about-us/san-roque-dam-a-legend-not-so-legend), The Bialet Masse Life Museum (https://bialetmasse.com/museobialetmasse/), and Wikipedia.

Historical photographs were also obtained from the Historical Archive of Cordoba, Ministry of Water, Environment, and Public Services of Cordoba Province.

The original design of the old San Roque Dam was developed also by French engineer Eugene Dumesnil, who worked alongside Casaffousth in the early stages of the project, before returning to France. While not featured in the comic, his role was fundamental in shaping the dam's initial concept.

FRIAS, L. R. (1985). "Historia del Dique San Roque". Cordoba Municipal Publishing House, 540 pp.

HUBER, N. (2001). "Paisaje y Vida del Valle Cordobes San Roque." Edicion Copiar Publishing.

Author´s notes:

My personal connection to this story stems from growing up near the shores of the San Roque Reservoir. My grandfather was often warned about the "danger" of the old dam during his stays in Cordoba city.

Although earlier hydraulic structures existed during pre-Hispanic and colonial times—such as small weirs and reservoirs in Peru, Bolivia, and Colombia—the San Roque Dam is often referred to as South America's first dam, as it was built using contemporary engineering principles. It marked the beginning of large-scale modern infrastructure in the region, incorporating industrial materials and European expertise.

This comic clearly shows how political interests can undermine scientists and engineers—a pattern that, sadly, persists in Argentina today, particularly in the water and environmental sectors.

The San Roque dam was more than an engineering feat—it was a battle between vision and resistance. It reshaped Cordoba's future for more than 50 years, until the creation of the new dam, securing water for a growing region. But those who built it paid a high price, facing political turmoil, accusations, and personal ruin.

Across the globe, another ambitious project was unfolding—one that would push the limits of engineering in an entirely different way. In the vast, arid heart of Western Australia, settlers and gold seekers faced a unique challenge: *water*.

Building a dam was one feat. But transporting life itself across 566 kilometers of unforgiving terrain? That was another.

It was an audacious mission—led by a man celebrated in Australia, yet unknown to much of the world.

- 4 -

THE GOLDEN PIPELINE

Engineering into the Outback

CHARLES YELVERTON O'CONNOR WAS BORN IN 1843 AT GRAVELMOUNT HOUSE, NEAR CASTLETOWN, IRELAND.
AT JUST 16, HE BECAME AN APPRENTICE UNDER THE SKILLED JOHN CHALONER SMITH, DIVING INTO THE WORLD OF RAILWAY ENGINEERING.
SMITH WASN'T JUST A CIVIL ENGINEER—HE ALSO WROTE ON BRITISH MEZZOTINTS.
BY 1864, AS IRELAND'S RAILWAY BOOM WAS SLOWING, IT WAS TIME FOR YOUNG O'CONNOR TO FIND NEW HORIZONS!
PEGASUS
ARMED WITH EXPERIENCE AND AMBITION, HE SET SAIL ON THE "PEGASUS" BOUND FOR NEW ZEALAND.
ONCE IN HIS NEW HOME, O'CONNOR ROLLED UP HIS SLEEVES: WATER FOR GOLD MINES, HARBORS, RAILWAYS, ROADS, BRIDGES...
HE BUILT THEM ALL!
HIS EXPERTISE EARNED HIM INTERNATIONAL RECOGNITION, AND IN 1880, HE WAS ELECTED TO THE INSTITUTION OF CIVIL ENGINEERS IN LONDON.
AFTER 25 YEARS OF EXCELLENCE IN NEW ZEALAND, O'CONNOR'S REPUTATION REACHED WESTERN AUSTRALIA—THEY NEEDED HIM!

SIR JOHN FORREST, WESTERN AUSTRALIA'S FIRST PREMIER, WASTED NO TIME. HE APPOINTED O'CONNOR AS ENGINEER-IN-CHIEF.
HE'D BE IN CHARGE OF RAILWAYS, HARBOURS, AND EVERYTHING!
BUT SOON, WESTERN AUSTRALIA'S GOLD RUSH ERUPTED, KICKING OFF IN 1885 WITH THE DISCOVERY OF AN 800-GRAM NUGGET IN THE NORTHEAST!
HALLS CREEK
EUREKA
OVER THE NEXT FEW YEARS, MORE GOLD WAS FOUND IN THE CENTRAL DESERTS OF WESTERN AUSTRALIA.
Kimberley
WESTERN AUSTRALIA
Kalgoorlie
Coolgardie
Perth
IN 1892, GOLD WAS STRUCK NEAR COOLGARDIE, AND IN 1894, IN KALGORLIE.
THOUSANDS RUSHED TO THE HARSH GOLDFIELDS, CHASING FORTUNE UNDER THE SCORCHING SUN.
COOLGARDIE'S POPULATION EXPLODED, BOOMING TO 15,000 IN NO TIME!
THE GOLD RUSH DEMANDED ACTION. A PROPER PORT WAS DESPERATELY NEEDED.
O'CONNOR ENVISIONED A DEEP INNER HARBOR AT THE MOUTH OF THE SWAN RIVER IN FREMANTLE, JUST 19 KM FROM PERTH.

EXPERT ENGINEERS SCOFFED: "THE HARBOUR IN THAT SITE WILL NEED CONSTANT DREDGING! IT'LL NEVER WORK!".
BUT O'CONNOR WAS DETERMINED, EVEN IF IT MEANT REMOVING A LIME BAR IN THE WAY.
THAT LIMESTONE BAR HAD TO GO!
Based on Photo 010722D, SLWA
YET, REMOVING THE BAR WOULD CHANGE THE RIVER'S SALINITY. IT THREATENED THE PLANTS AND FISH THE NOONGAR PEOPLE RELIED ON—FOR FOOD AND FOR CULTURE.
FOR THE NOONGAR, THE RIVER WAS SACRED—A PLACE OF SPIRIT AND TRADITION.
No photo references were used in the creation of this image.
DESPITE OPPOSITION, FUNDING WAS SECURED IN LONDON, AND BY LATE 1892, BLASTING AND DREDGING BEGAN.
AT LAST, LARGER SHIPS COULD SAFELY SAIL INTO THE HARBOR, BRINGING PROSPERITY TO WESTERN AUSTRALIA.
Based on Photo 009026D, SLWA.
BUT THE HARBOR WASN'T O'CONNOR'S ONLY CHALLENGE. OUT IN THE GOLDFIELDS, A CRISIS LOOMED.
GOLD BROUGHT WEALTH TO WESTERN AUSTRALIA, BUT THE GOLDFIELDS REMAINED PARCHED, A VAST DESERT OF DUST AND DESPERATION. THOUSANDS STRUGGLED TO SURVIVE WITHOUT WATER.
WATER COSTS MORE THAN WHISKEY!
ARTESIAN BORES AND CONDENSING PLANTS HELPED, BUT WEREN'T ENOUGH.

NO FRESH WATER MEANT POOR SANITATION AND DEADLY DISEASES LIKE TYPHOID. MANY ALSO DIED FROM "CONDENSER SICKNESS"—CAUSED BY DRINKING WATER DISTILLED WITH WOOD-FIRED CONDENSING PLANTS.
BY 1891, CEMETERIES IN THE GOLDFIELDS WERE FILLING UP FAR TOO QUICKLY—EACH GRAVE MARKED ANOTHER LIFE CUT SHORT.
WHILE PERTH ENJOYED OVER 700MM OF RAIN A YEAR,
THE GOLDFIELDS BARELY SAW 230MM —YEAR AFTER YEAR.
SIR FORREST KNEW THAT FOR THE GOLDFIELDS TO THRIVE, LUCK WASN'T ENOUGH. THEY NEEDED WATER—AND A BOLD PLAN TO GET IT.
THESE PEOPLE NEED A LIFELINE, O'CONNOR!
IN 1893, O'CONNOR PROPOSED HIS MOST DARING IDEA YET: A PIPELINE TO BRING WATER FROM PERTH TO THE DESERT!
WHY NOT PUMP ALL THAT UNUSED COASTAL WATER UP TO THE GOLDFIELDS?
THE LONGEST PIPELINE EVER BUILT! LEAKS WILL WASTE ALL THE WATER!
WHO COULD BELIEVE THIS IS POSSIBLE?
700mm
200mm
GOLDFIELDS
pipeline
PERTH
A 566 KM PIPELINE—CLIMBING 390 M WITH 8 STEAM PUMPS—WOULD DELIVER 23,000 CUBIC METERS OF WATER A DAY!
MANY CALLED THE IDEA MADNESS. BUT O'CONNOR KNEW: IT WAS THE ONLY WAY TO SUSTAIN THE GOLDFIELDS.

HE IGNORED THE DOUBTERS. THREW HIMSELF INTO SURVEYING, LEVELING, AND COMPENSATING SETTLERS—LAND THAT WOULD SOON HOLD PERTH'S MASSIVE NEW DAM: "MUNDARING WEIR".
Mt. Charlotte Reservoir (Kalrgoorlie)
566 km
Mundaring Weir Reservoir (Perth)
TO STREAMLINE THE JOURNEY, HE ALIGNED THE PIPELINE WITH THE RAILWAY, WHICH HAD FINALLY REACHED THE GOLDFIELDS IN EARLY 1897.
SOME EXPERTS BACKED HIM—ENGINEER T.C. HODGSON PUT IT SIMPLY: "WE'VE PUMPED WATER TO TANKS BEFORE—NOW WE JUST NEED TO DO IT EIGHT TIMES IN A ROW".
700mm
200mm
GOLDFIELDS
pipeline
PERTH
WESTERN AUSTRALIA TO THE GOLOFIELDS
8 X
HODGSON EVEN CHALLENGED THE PLAN, PROPOSING LARGER PIPES AND MORE POWERFUL PUMPS!
CRITICS ERUPTED WHEN THEY HEARD THE COST—THIS PIPELINE WOULD MATCH WESTERN AUSTRALIA'S ENTIRE ANNUAL BUDGET!
IT'S MADNESS! MORTGAGING OUR FUTURE FOR A SINGLE PROJECT?
TWO AND A HALF MILLION POUNDS ON OUR SMALL COMMUNITY'S SHOULDERS?
NO GOVERNMENT CAN JUSTIFY PLEDGING THE COUNTRY'S CREDIT FOR WATER SUPPLY TO THE MINES!
BUT PREMIER SIR FORREST HAD NO DOUBTS. HE PUSHED THE PROJECT THROUGH PARLIAMENT AND SECURED A LOAN FROM ENGLAND.

Based on Photo 007130D, SLWA

Based on Photo 000810D, SLWA

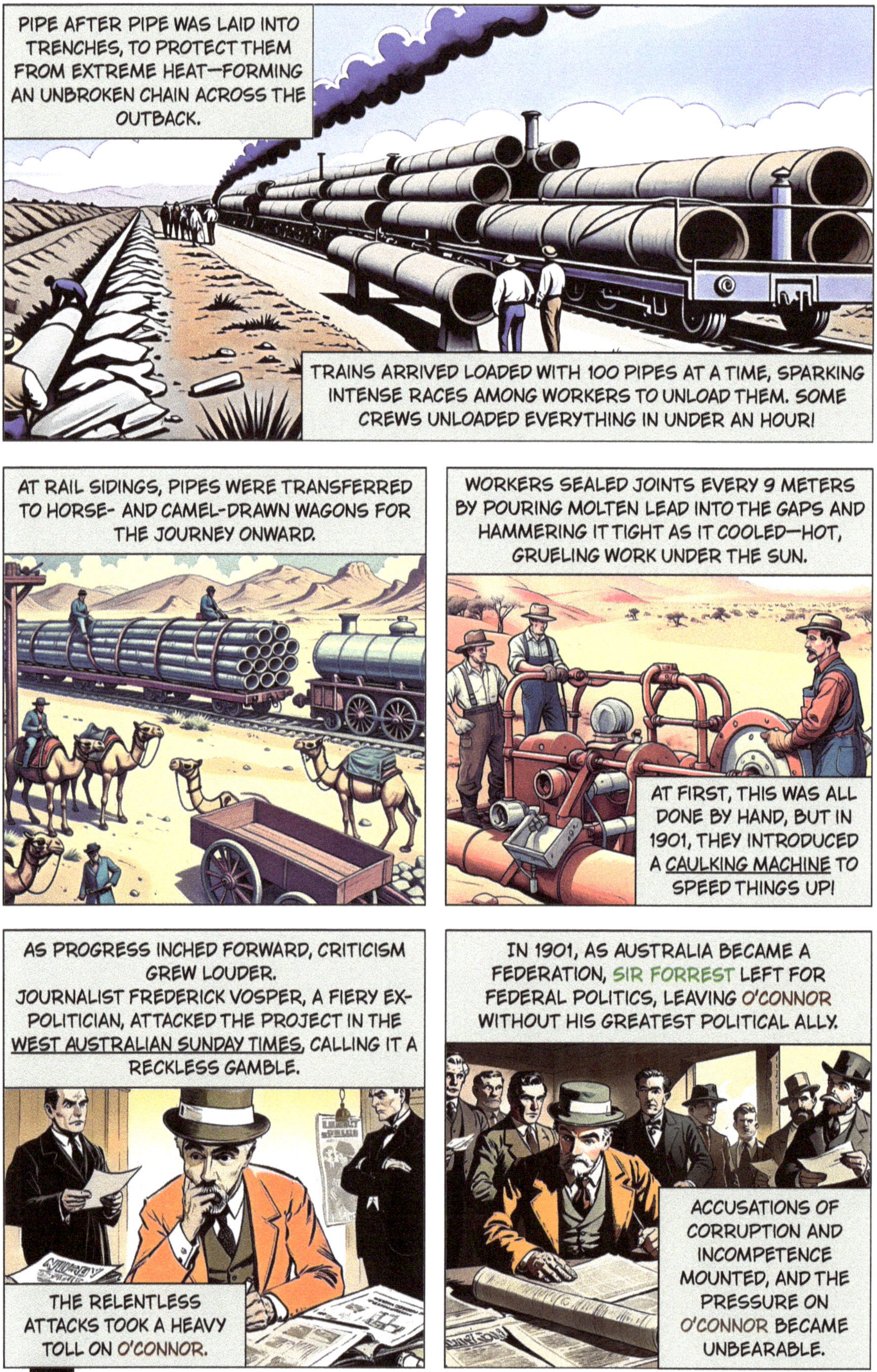

Based on Photo 0008334D, SLWA

IN EARLY 1902, DEBATES OVER PIPELINE DELAYS AND CONTROVERSIAL LAND DEALS TRIGGERED A GOVERNMENT INQUIRY, ADDING FUEL TO THE FIRE.
PIPE CAULKING
AND LAND DEALICTS
ALONG A PIPELINE ROUTE
ON MARCH 10, 1902, DESPITE SUCCESS ON THE GROUND—STATION #1 WAS READY, 12 KM TESTED—O'CONNOR WROTE HIS FINAL LETTER.
"THE SITUATION HAS BECOME IMPOSSIBLE".
THAT MONDAY MORNING, HE QUIETLY SADDLED HIS HORSE AND LEFT HIS EAST FREMANTLE HOME—FOR THE LAST TIME.
HE USUALLY RODE ALONG THE SHORE WITH HIS DAUGHTER BRIDGET, BUT THAT DAY, HE WENT ALONE—SHE WASN'T FEELING WELL.
AS HE RODE INTO THE SEA, O'CONNOR TOOK HIS OWN LIFE, CRUSHED BY THE WEIGHT OF RELENTLESS CRITICISM AND DOUBT.
BY 7:30 A.M., A BOY FOUND A RIDERLESS HORSE NEAR ROBB'S JETTY. FOLLOWING ITS TRACKS TO THE SHORE, HE SAW A BODY IN THE WAVES—CHARLES YELVERTON O'CONNOR WAS GONE.
TWO MONTHS LATER, THE INQUIRY CLEARED O'CONNOR—THERE WAS NO BASIS FOR THE ACCUSATIONS.
THESE ACCUSATIONS ARE BASELESS!
ENGINEER HODGSON, HIS DEPUTY, FACED NO CHARGES BUT RESIGNED SHORTLY AFTER.

Based on Photo 013543D, SLWA

Based on Photo 229080PD, SLWA

Source: National Trust Western Australia.

IN THE 1930S, IT WAS ALSO RAISED ABOVE GROUND FOR EASIER MAINTENANCE—JUST AS IT APPEARS TODAY.
ONCE THE WORLD'S LONGEST PIPELINE, IT REMAINS THE LONGEST STEEL PIPELINE STILL IN USE —A TESTAMENT TO ITS ENDURING DESIGN.

Kalgoorlie, modern view from Mt. Charlotte Reservoir.

Super-Pit gold mine located near Kalgoorlie-Boulder.

Horse and Rider Monument at C.Y. O'Connor Beach, commemorating O'Connor's death on the site in 1902.

THE GOLDFIELD PIPELINE TODAY

The Coolgardie Goldfields Water Supply Scheme was designated as a **National Engineering Landmark** by Engineers Australia in recognition of its historical and engineering significance.

In 2009 was also recognized as an **International Historic Civil Engineering Landmark** by the American Society of Civil Engineers.

A PIONEERING FEAT OF WATER TRANSFER AND ENGINEERING

The Goldfields pipeline is recognised as the first modern example of **interbasinal transfer of water.**

It was one of the first major projects to use steel piping on such a massive scale—setting the standard for future infrastructure.

Statue of C Y O'Connor at Freemantle, Western Australia.

Sources & References

For this documentary comic, I drew on several key sources I became familiar with while living in Australia. During my PhD studies in Perth, I conducted research at the reservoir formed by Mundaring Weir—a key component of the water pipeline scheme featured in the comic. This reservoir was officially renamed Lake C.Y. O'Connor in 2004.

The National Trust of Western Australia's Golden Pipeline project (goldenpipeline.com.au) offered detailed historical insights and engineering background that helped shape the narrative.

The Western Australian Museum (museum.wa.gov.au) contributed valuable archival material, while historical photographs—used with permission from the State Library of Western Australia (SLWA)—ensured visual accuracy throughout many scenes.

The image of the semi-submerged Horse and Rider statue commemorating O'Connor's death was taken by Nachoman-au and is available on Wikipedia:
(https://commons.wikimedia.org/wiki/File:C.Y._O%27Connor_in_the_surf.jpg).

O'Connor's pipeline had a profound impact on the Australian landscape. It didn't just bring water to an arid land—it fueled a gold rush, transforming Western Australia into the nation's largest gold-producing state. Water was no longer just a necessity; it became the force that drove prosperity and expansion.

Meanwhile, across the globe, another monumental dream was taking shape—this time not in the service of a single nation, but as part of a grander vision: a channel that would reshape global trade and project imperial ambition.

In the jungles of Panama, an audacious plan was emerging—to carve a path between oceans. Explorers had long dreamed of it, merchants longed for it—and now, fresh from his triumph at the Suez Canal, *Ferdinand de Lesseps* was setting out to make it real.

A canal through Panama would change the world—but at what cost? What will it take to turn an imperial dream into reality?

- 5 -

EMPIRE BY DESIGN

Carving a Path Between the Oceans

JANUARY 10, 1880—THE JUNGLE TREMBLED!
BOOM!
TODAY, WE BEGIN THE GREATEST ENGINEERING FEAT OF OUR AGE! MORE THAN A CANAL—IT WILL BE A GIFT TO PROGRESS, TO ALL MANKIND.
THAT SYMBOLIC BLAST ECHOED FAR BEYOND PANAMA—CARRYING WITH IT AMBITION, EMPIRE, AND THE WEIGHT OF A DREAM THAT HAD JUST BEEN SET IN MOTION.
FERDINAND DE LESSEPS STOOD PROUDLY BESIDE HIS DAUGHTER, WATCHING HISTORY BEGIN. HIS BOLD NEW DREAM: CARVE A PATH BETWEEN OCEANS—AND RESHAPE THE WORLD.

DE LESSEPS—FRANCE'S HERO, THE "CONQUEROR OF THE DESERT"—WAS THE STAR OF THE MOMENT!
A DECADE EARLIER, IN 1869, HE HAD UNVEILED THE SUEZ CANAL, LINKING THE MEDITERRANEAN AND RED SEAS.

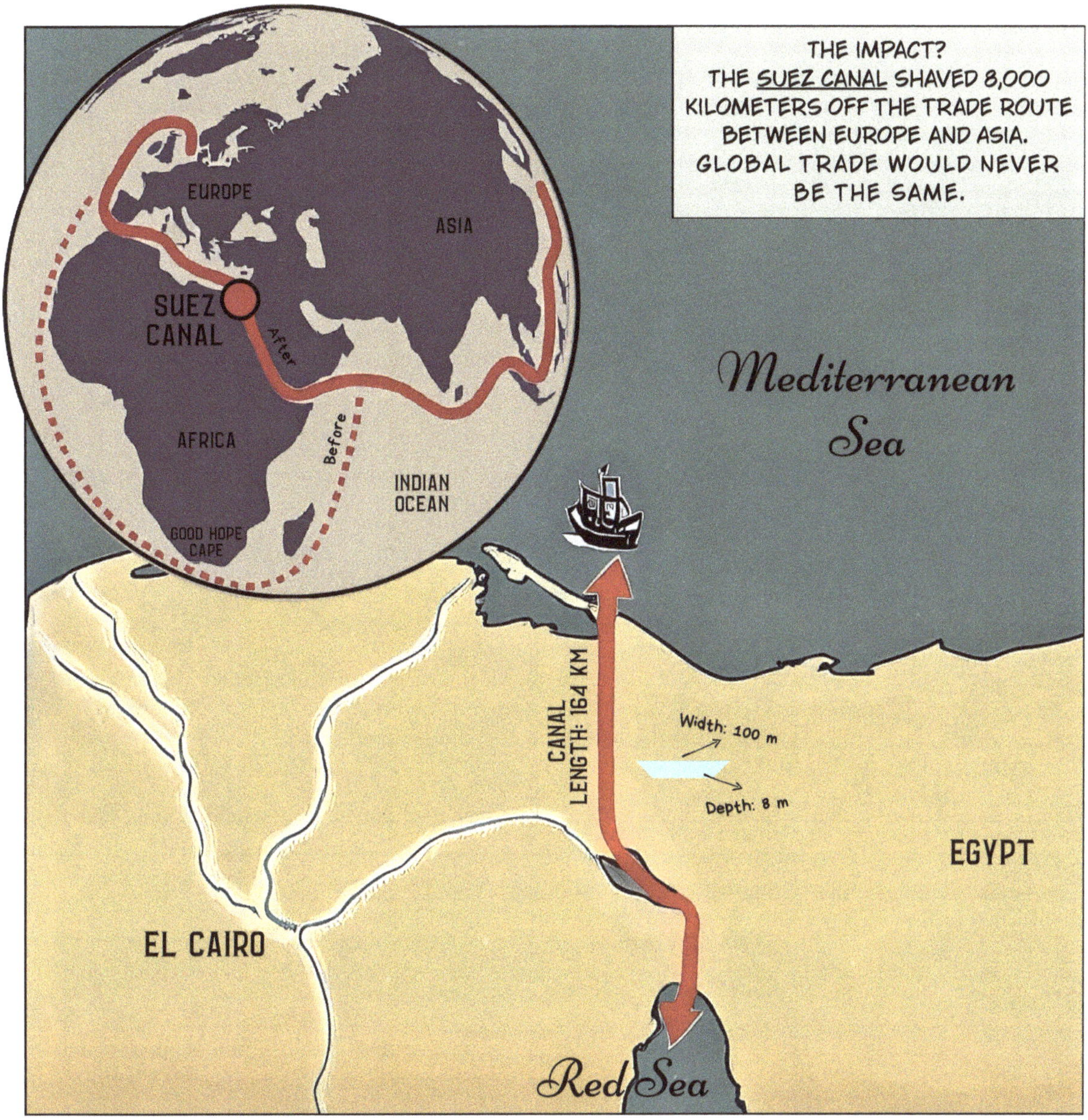
THE IMPACT?
THE SUEZ CANAL SHAVED 8,000 KILOMETERS OFF THE TRADE ROUTE BETWEEN EUROPE AND ASIA. GLOBAL TRADE WOULD NEVER BE THE SAME.
EUROPE
ASIA
SUEZ CANAL
After
AFRICA
Before
INDIAN OCEAN
GOOD HOPE CAPE
Mediterranean Sea
CANAL LENGTH: 164 KM
Width: 100 m
Depth: 8 m
EGYPT
EL CAIRO
Red Sea

NOW, ON THE OTHER SIDE OF THE WORLD, DE LESSEPS DREAMED AGAIN: CUT THROUGH THE JUNGLE, CARVE A CANAL THROUGH THE HEART OF PANAMA ISTHMUS... AND UNITE THE ATLANTIC AND PACIFIC OCEANS.

BUT DE LESSEPS WASN'T THE FIRST TO DREAM OF UNITING THE OCEANS. IN 1513, SPANISH EXPLORER VASCO NÚÑEZ DE BALBOA REACHED PANAMA'S ATLANTIC COAST.
ARE YOU SAYING THERE'S ANOTHER SEA?
LOCALS SPOKE OF "ANOTHER SEA"—VAST, MYSTERIOUS... THE PACIFIC.
INTRIGUED, BALBOA CROSSED THE 110-KILOMETER ISTHMUS AND BECAME THE FIRST EUROPEAN TO LAY EYES ON THE PACIFIC—THE "SOUTH SEA," AS HE CLAIMED IT FOR SPAIN.
BALBOA'S DISCOVERY TRANSFORMED PANAMA INTO A VITAL TRADE AND MILITARY CORRIDOR.
BY THE 1500S, MULE TRAINS CARRIED GOLD AND SILVER ACROSS THE ISTHMUS, FORGING AN INVISIBLE ROAD OF EMPIRE THROUGH THE JUNGLE.

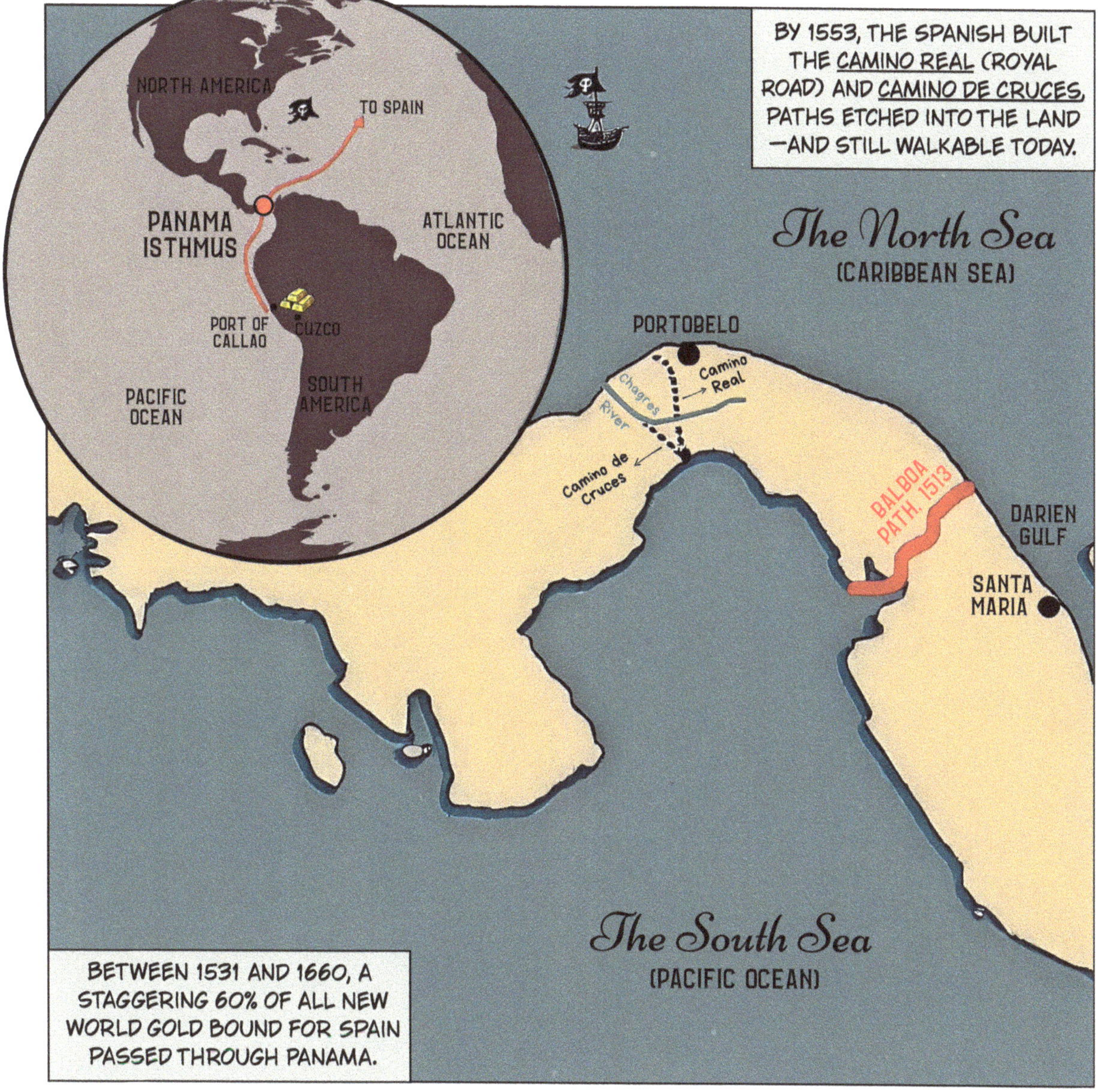

BETWEEN 1606 AND 1739, "THE PORTOBELO FAIRS" FLOURISHED—MASSIVE TRADE FILLED THE AMERICAS WITH EUROPEAN GOODS (FOOD, CERAMICS, TEXTILES) FUELING COLONIAL ECONOMIES AND OVERFLOWING THE PORTS WITH ACTIVITY.

BY 1520, EMPEROR CHARLES V (KING CHARLES I OF SPAIN) HAD ALREADY ENVISIONED A CANAL. HIS ENGINEERS EXPLORED POSSIBLE ROUTES... BUT THE DREAM REMAINED OUT OF REACH.
IT'S AN IMPOSSIBLE FEAT!
PANAMA
DURING THE 18TH CENTURY, MORE PLANS AND REPORTS EMERGED, BUT THE SPANISH CROWN PRIORITIZED OTHER VENTURES.

HIS SON, KING PHILIP II ("THE PRUDENT"), LATER ABANDONED THE IDEA, FEARING THE 20 CM SEA LEVEL DIFFERENCE BETWEEN OCEANS WOULD FLOOD THE ISTHMUS.
TOO RISKY!

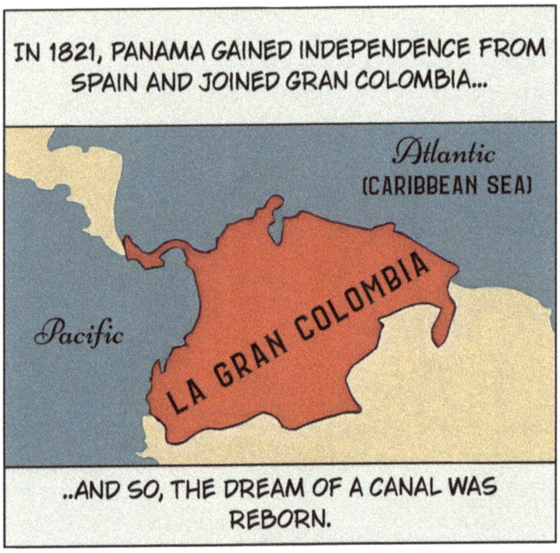
IN 1821, PANAMA GAINED INDEPENDENCE FROM SPAIN AND JOINED GRAN COLOMBIA...
Atlantic
(CARIBBEAN SEA)
Pacific
LA GRAN COLOMBIA
..AND SO, THE DREAM OF A CANAL WAS REBORN.

BY 1849, THE CALIFORNIA GOLD RUSH MADE PANAMA A KEY TRANSIT ROUTE. GOLD-SEEKERS CROSSED BY MULE TO REACH THE PACIFIC, THEN CONTINUED BY SEA TO THE U.S.

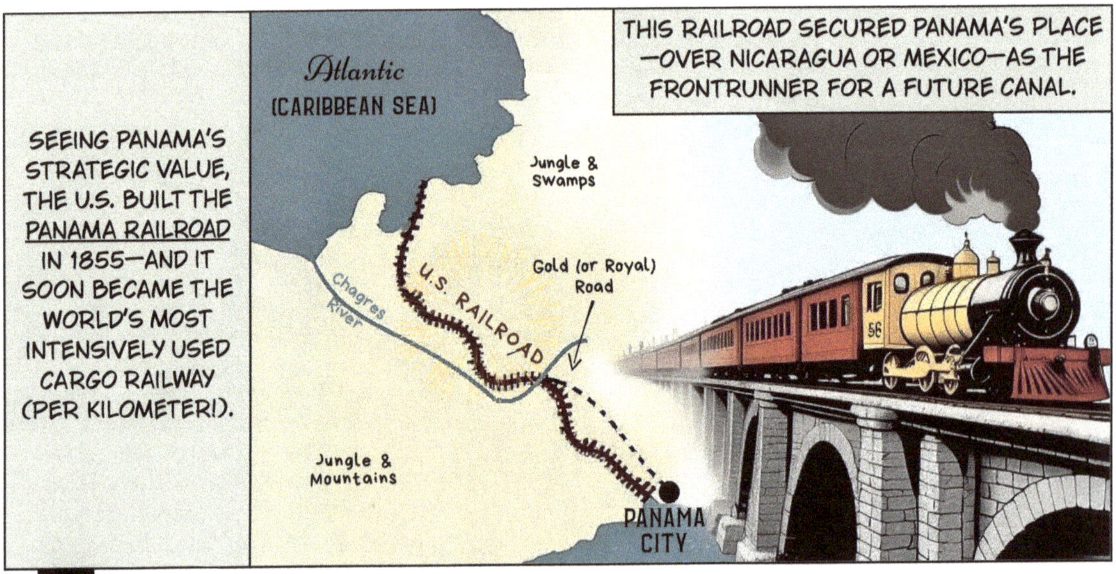
SEEING PANAMA'S STRATEGIC VALUE, THE U.S. BUILT THE PANAMA RAILROAD IN 1855—AND IT SOON BECAME THE WORLD'S MOST INTENSIVELY USED CARGO RAILWAY (PER KILOMETER!).
Atlantic
(CARIBBEAN SEA)
Jungle & Swamps
Chagres River
U.S. RAILROAD
Gold (or Royal) Road
Jungle & Mountains
PANAMA CITY
THIS RAILROAD SECURED PANAMA'S PLACE —OVER NICARAGUA OR MEXICO—AS THE FRONTRUNNER FOR A FUTURE CANAL.
56

IN 1877, DE LESSEPS STEPPED BACK INTO THE SPOTLIGHT. FRESH FROM HIS TRIUMPH AT THE SUEZ CANAL, HE SECURED THE RIGHTS TO BUILD A CANAL THROUGH PANAMA. HIS ENVOY, FRENCH ENGINEER LUCIEN NAPOLEON BONAPARTE WYSE, PROPOSED A LOCK-BASED DESIGN.
LOCKS? NONSENSE! SUEZ NEEDED NONE —AND NEITHER WILL PANAMA!
BUT DE LESSEPS—NO ENGINEER HIMSELF—UNDERESTIMATED PANAMA'S BRUTAL TERRAIN. HE REJECTED WYSE'S PLAN, DETERMINED TO BUILD A SEA-LEVEL CANAL, JUST LIKE SUEZ.
CANAL LOCKS WORK LIKE WATER ELEVATORS, LIFTING AND LOWERING SHIPS TO OVERCOME ELEVATION CHANGES.
LOCK CHAMBERS—THE "ELEVATORS" OF THE CANAL—ARE FILLED (OR EMPTIED) WITH WATER TO RAISE OR LOWER SHIPS.
STILL, DE LESSEPS CLUNG TO HIS SEA-LEVEL PLAN—ONE THAT DEMANDED CARVING THROUGH THE GREATEST OBSTACLE OF ALL: "CULEBRA" MOUNTAIN.
A SEA-LEVEL CANAL, JUST LIKE SUEZ!
PANAMA CANAU

IN 1878, THE "WYSE TREATY" (SIGNED IN BOGOTA) GRANTED FRANCE A 99-YEAR CONCESSION TO BUILD AND OPERATE THE CANAL.
WYSE TREATY
AFTER THAT, CONTROL WOULD RETURN TO COLOMBIA—PANAMA'S GOVERNING NATION AT THE TIME.
IN 1879, A GLOBAL CONGRESS CONVENED IN PARIS.
DELEGATES FROM 25 COUNTRIES—NCLUDING FRANCE, SPAIN, THE U.S., AND MEXICO—GATHERED TO MAKE HISTORY.
THE WORLD WATCHED IN AWE... IT WAS DECLARED THAT THE UNIVERSAL COMPANY OF THE INTEROCEANIC CANAL OF PANAMA, LED BY DE LESSEPS, WOULD TAKE CHARGE OF THE MONUMENTAL PROJECT.
WITH FUNDING SECURED, DE LESSEPS BOLDLY REVEALED HIS VISION: A SEA-LEVEL CANAL, CARVED STRAIGHT THROUGH PANAMA — JUST LIKE SUEZ.
A CANAL IN PANAMA WOULD BE A TRIUMPH FOR FRANCE, AND A NEW GATEWAY FOR GLOBAL TRADE!
IN PARIS, THE CANAL WASN'T MERELY A PROJECT—IT WAS A PROMISE: OF GLORY, COMMERCE, AND CONTROL.
BETWEEN 1871 AND 1881, DE LESSEPS LED THE PARIS GEOGRAPHICAL SOCIETY, RALLYING SUPPORT FOR HIS PROPOSAL OF A SEA-LEVEL CANAL IN PANAMA.

Come, Lesseps! At the sound of your name,
The Isthmus trembles.Humanity awaits you.
Here, a new victory,
A new laurel will add to your glory.
The whole world supports you.
Come, go forth, and conquer. Hurry now...
Your prodigious genius
Will transform into reality with your resolve
What inspired men once envisioned,
What human progress has long desired,
What science was meant to prove,
What God's hand itself had pointed to.
Even if the rocks were made of diamonds,
Soon you will see your dreams fulfilled.
Come: as you once united two pygmies,
Now you shall come to unite two giants.
HAILED AS A HERO OF HIS TIME, DE LESSEPS WAS WIDELY PRAISED FOR HIS VISION.
THE SPANISH POET MARÍA GUTIÉRREZ DE ALBA CALLED HIM "THE GENIUS OF INGENIOUS", CELEBRATING HIS AMBITION.

IN 1880, CONSTRUCTION OF THE LONG-AWAITED PANAMA CANAL FINALLY BEGAN!
BUT SOON, REALITY STRUCK: PANAMA WAS NO EGYPT. IT WAS WILDER, DEADLIER, AND FAR LESS FORGIVING.

FIRST, THE JUNGLE FOUGHT BACK!
TROPICAL DISEASES—MALARIA, YELLOW FEVER, CHOLERA—RAVAGED THE WORKFORCE.
THEN CAME MISMANAGEMENT. TOGETHER, DISEASE AND MISMANAGEMENT DRAINED LIVES AND RESOURCES—THE PROJECT BEGAN TO CRUMBLE.

IN 1882, THE EARTH ITSELF REBELLED! A 7.9-MAGNITUDE QUAKE—THE "GREAT EARTHQUAKE"—SHOOK THE PANAMA ISTHMUS.
CANAL CONSTRUCTION HALTED TEMPORARILY.

BY 1883, THE CRACKS WEREN'T JUST PHYSICAL. EXCAVATION TURNED CHAOTIC—EARLY STUDIES UNDERESTIMATED THE TERRAIN.
?
EXCAVATION WORK

YET, IN THE FACE OF DISASTER, THE DIGGING CONTINUED.
BY 1884, OVER 19,000 WORKERS HAD COME FROM ACROSS THE CARIBBEAN, BRINGING STRENGTH—AND SACRIFICE.
AT ITS PEAK, THE FRENCH PROJECT EMPLOYED AN ESTIMATED 40,000 WORKERS FROM VARIOUS REGIONS.

IN 1885, YELLOW FEVER AND MALARIA STRUCK AGAIN—THIS TIME WITH EVEN GREATER FORCE, CLAIMING TOP PROJECT DIRECTORS.
AT THE TIME, TROPICAL DISEASES WERE MISUNDERSTOOD. DOCTORS BLAMED KIDNEY FAILURE, POISONED AIR AND SOIL—NEVER SUSPECTING MOSQUITOES.

THE GENERAL DIRECTOR OF THE PROJECT, JULES DINGLER, LOST HIS SON, HIS DAUGHTER AND HER FIANCE—AND THEN HIS WIFE, TO YELLOW FEVER.

BROKEN AND ALONE, HE LEFT PANAMA IN 1885, NEVER TO RETURN. GRIEF CONSUMED HIM. HE DIED SHORTLY AFTER.
BETWEEN 1883-1888 SIX DIFFERENT GENERAL DIRECTORS STRUGGLED TO HOLD THE PROJECT TOGETHER.

BY 1886, LANDSLIDES HAD BECOME YET ANOTHER DEADLY ENEMY.
UNSTABLE HILLSIDES CRUMBLED INTO THE CANAL, FORCING CREWS TO DIG THE SAME SECTIONS AGAIN AND AGAIN—ONLY TO WATCH THEIR PROGRESS VANISH IN SECONDS.
LAND RESISTED. EQUIPMENT FAILED. THE DISEASES SPREAD. AND YET, IN PARIS, THE BROCHURES STILL SPOKE OF TRIUMPH.

FIVE YEARS INTO WHAT WAS PLANNED AS AN EIGHT-YEAR PROJECT, ONLY 10% OF THE "CULEBRA CUT"—THE MOST DIFFICULT SECTION—WAS COMPLETED!
THEY THOUGHT THEY COULD CUT THE LAND LIKE PAPER.
WAS A SEA-LEVEL CANAL EVEN POSSIBLE? DOUBTS WERE NO LONGER A WHISPER—THEY ECHOED THROUGH EVERY CAMP.
BY THE END OF THE FRENCH PROJECT, OVER 28,000 WORKERS WILL HAVE LOST THEIR LIVES—TO DISEASE AND DEADLY LANDSLIDES.

GENERAL DIRECTOR BOYER ADMITTED THE SEA-LEVEL CANAL WAS UNFEASIBLE. HE PROPOSED A LOCK-BASED SYSTEM WITH AN ARTIFICIAL LAKE INSTEAD.
NOT WITHOUT LOCKS!
BUT DE LESSEPS REFUSED TO LISTEN!
BY 1887, PRESSURE MOUNTED. POLITICIANS, INVESTORS, AND THE PUBLIC DEMANDED ANSWERS.
DEFEATED BY THE HARSH REALITY, DE LESSEPS FINALLY AGREED TO A LOCK-BASED CANAL AND HANDED THE PROJECT OVER TO FRENCH ENGINEER, GUSTAVE EIFFEL.
BY THAT TIME, EIFFEL WAS ALREADY A LEGEND, WITH ICONIC DESIGNS AROUND THE WORLD...
THE FRAME FOR THE STATUE OF LIBERTY
THE MARIA PIA BRIDGE, PORTUGAL
THE GARABIT VIADUCT, FRANCE
THE DOME OF THE NICE OBSERVATORY, FRANCE
...AND HIS MOST DARING CREATION—THE EIFFEL TOWER—WAS RISING IN PARIS...
WORK ON THE LOCK CANAL BEGAN ON JANUARY 15, 1888.
EIFFEL INTRODUCED VISIONARY INNOVATIONS:
✓ MOVABLE SLUICE GATES (SLIDING GATES)
✓ MULTIPLE WATER CHANNELS TO IMPROVE FLOW EFFICIENCY.

THAT YEAR, CONSTRUCTION SURGED FORWARD! THE FIRST OF TEN LOCKS WAS UNDERWAY.
BUT BEHIND THE SCENES, CORRUPTION RAN DEEP:
◆ DUPLICATE CONTRACTS DRAINED FUNDS.
◆ OFFICIALS LIVED LARGE WHILE WORKERS SUFFERED.
◆ ABSURD PURCHASES—100 SNOWPLOWS... FOR A SNOWLESS REGION?!
JUST AS PROGRESS GAINED MOMENTUM... THE MONEY VANISHED.
DEBTS
IN DESPERATION, DE LESSEPS' CIRCLE BRIBED JOURNALISTS TO HIDE THE CRISIS AND EVEN LAUNCHED A LOTTERY TO RAISE FUNDS.
LOTERIE
IT FAILED.
BY 1889, THE SCANDAL EXPLODES! FRENCH NEWSPAPERS REVEALED THE TRUTH: FRAUD, EMBEZZLEMENT, AND BRIBES INVOLVING OVER 100 POLITICIANS.
PANAMA CANAL
FRAUD!
PANAMA CANAL EXPOSED!
A MASSIVE CORRUPTION SCANDAL WAS UNFOLDING—ONE OF THE CENTURY'S WORST.

INVESTIGATIONS UNCOVERED MASSIVE FRAUD. DE LESSEPS, HIS SON CHARLES, AND EVEN EIFFEL WERE CONVICTED OF MISMANAGEMENT!
IT WAS AN UNWORTHY END FOR SUCH A GRAND EFFORT.

BARON REINACH, FINANCIAL OVERSEER OF THE INTEROCEANIC CANAL COMPANY, COMMITTED SUICIDE, LIKELY AFTER DESTROYING EVIDENCE.
DE LESSEPS WAS SENTENCED—THEN CLEARED FOR LACK OF EVIDENCE. HE DIED SHORTLY AFTER THE TRIALS, IN 1894 AT AGE 89. CHARLES, HIS SON, WAS CONVICTED AND BRIEFLY IMPRISONED.

I'M STEPPING DOWN.
EIFFEL FOUGHT BACK. HIS CONVICTION WAS OVERTURNED—NO EVIDENCE HAD EVER LINKED HIM DIRECTLY.
BEFORE THE TRIAL, HE HAD ALREADY RESIGNED FROM HIS COMPANY, "ETABLISSEMENTS EIFFEL", AND DEMANDED THAT HIS NAME BE REMOVED.
THE DAMAGE WAS DONE. EIFFEL WALKED AWAY FROM CONSTRUCTION FOREVER... AND TURNED TO SCIENCE INSTEAD.

FOR THE NEXT 30 YEARS, EIFFEL DEDICATED HIMSELF TO SCIENTIFIC RESEARCH.
HE FOCUSED ON FINDING A PURPOSE FOR THE EIFFEL TOWER, WHICH HAD INITIALLY BUILT WITH A 20-YEAR LIFESPAN. HE TURNED THE TOWER INTO A HUB FOR EXPERIMENTS IN AERODYNAMICS, METEOROLOGY, AND RADIO COMMUNICATION.
IN THE WWI THAT WAS YET TO COME, THE EIFFEL TOWER WOULD SERVE AS A KEY ANTENNA FOR ALLIED ESPIONAGE.

THE FRENCH HAD FAILED, BUT THEIR EFFORTS WEREN'T IN VAIN. THEY LEFT BEHIND EXCAVATION RECORDS, CLINICS, AND KEY ENGINEERING INSIGHTS.

IN 1902, U.S. PRESIDENT ROOSEVELT SIGNED THE "SPOONER ACT", AUTHORIZING THE PURCHASE OF FRENCH CANAL ASSETS FOR $40 MILLION.
IT WAS THE END OF ONE EMPIRE'S DREAM... AND THE BEGINNING OF ANOTHER'S.

IN 1903, PANAMA, WITH U.S. BACKING, DECLARED INDEPENDENCE FROM COLOMBIA, CLEARING THE POLITICAL OBSTACLE TO CANAL CONSTRUCTION.
Central America
Atlantic (Caribbean sea)
Pacific
South America
COLOMBIA'S GOVERNMENT HAD ALWAYS BEEN RELUCTANT TO ALLOW FOREIGN INFLUENCE IN SUCH A CRUCIAL AND STRATEGIC REGION.
IN 1904, THE CANAL GOT A SECOND CHANCE AT SUCCESS. THE U.S. OFFICIALLY TOOK OVER AND REVIVED THE PROJECT.
LET'S FINISH THIS CANAL ONCE AND FOR ALL!
AT LAST, THE KEY TO STOPPING YELLOW FEVER WASN'T MEDICINE—IT WAS MOSQUITOES! TO STOP THEM, DR. WILLIAMS GORGAS LED A REVOLUTIONARY CAMPAIGN:
✓ DRAINED STAGNANT WATER
✓ FUMIGATED HOMES
✓ ENFORCED QUARANTINES
CUBAN DR. CARLOS FINLAY HAD DISCOVERED IN 1881 THAT MOSQUITOES SPREAD YELLOW FEVER—BUT HIS WORK WAS IGNORED UNTIL 1900!
BY 1906, YELLOW FEVER WAS FINALLY ERADICATED IN THE CANAL ZONE, ALLOWING CONSTRUCTION TO MOVE FORWARD.

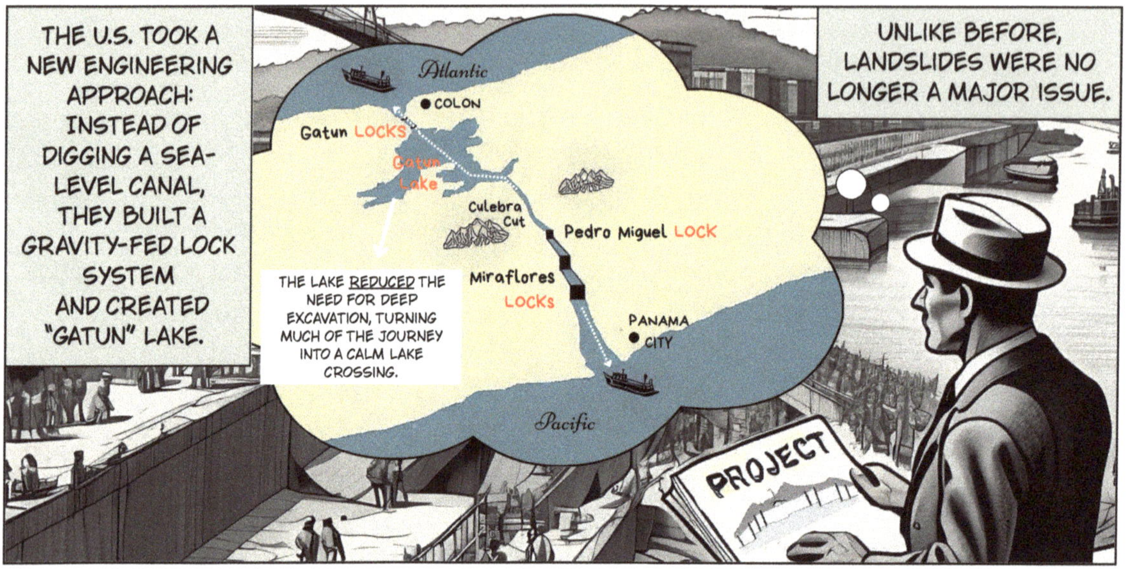
THE U.S. TOOK A NEW ENGINEERING APPROACH: INSTEAD OF DIGGING A SEA-LEVEL CANAL, THEY BUILT A GRAVITY-FED LOCK SYSTEM AND CREATED "GATUN" LAKE.
Atlantic
COLON
Gatun LOCKS
Gatun Lake
Culebra Cut
Pedro Miguel LOCK
Miraflores LOCKS
PANAMA CITY
Pacific
THE LAKE REDUCED THE NEED FOR DEEP EXCAVATION, TURNING MUCH OF THE JOURNEY INTO A CALM LAKE CROSSING.
UNLIKE BEFORE, LANDSLIDES WERE NO LONGER A MAJOR ISSUE.
PROJECT

ON AUGUST 15, 1914, THE PANAMA CANAL OPENED!. 77 KM OF ENGINEERING BRILLIANCE THAT FOREVER CHANGED GLOBAL TRADE, MARKED BY THE HISTORIC FIRST CROSSING OF THE STEAMSHIP "ANCON".
MORE THAN 42,000 WORKERS CONTRIBUTED DURING THE U.S. PHASE. SADLY, 5,600 LIVES WERE LOST.
ANCON
ANCON
THE U.S. CONTROLLED THE CANAL UNTIL 1999, WHEN PANAMA ASSUMED FULL OWNERSHIP.
THE ORIGINAL LOCKS WERE 34 METERS WIDE—BUT IN 2016, NEW EXPANDED LOCKS OPENED, ALLOWING SHIPS TWICE AS LARGE!
MORE THAN A CENTURY AFTER ITS OPENING, THE CANAL REMAINS A VITAL ARTERY FOR GLOBAL COMMERCE—BORN FROM AMBITION, CARVED THROUGH JUNGLE, LIFTED BY SACRIFICE, AND CARRIED FORWARD BY GENERATIONS. TODAY, IT MOVES 3% OF GLOBAL TRADE!

CIVIL ENGINEERING MONUMENT OF THE MILLENNIUM

In 1994, the American Society of Civil Engineers (ASCE) recognized the Panama Canal as one of the **Seven Wonders of the Modern World** due to its groundbreaking engineering and global impact.

The ASCE also recognized the canal as a **National Historic Civil Engineering Landmark** due to its significance in global trade and engineering innovation.

A look down the Miraflores Locks.

A GLOBAL ENGINEERING MARVEL

The Locks System and the Expansion Project (completed in 2016) have been recognized globally for their technological advancements, including water-saving basins and larger lock chambers.

PORTOBELO-SAN LORENZO: UNESCO HERITAGE SITE

In 1980, UNESCO recognized the Portobelo-San Lorenzo Caribbean fortifications as a **World Heritage Site**, acknowledging their architectural brilliance and strategic importance along the historic trade route linked to the Panama Canal.

A ship enters the Miraflores Locks, guided by electric locomotives—still operating since 1914.

Sources & References

This documentary comic is based on the history of the canal as presented by the Autoridad del Canal de Panama (pancanal.com). For a contemporary perspective on its construction—covering technical challenges and key engineers, and geopolitical impact—I also drew from Ira E. Bennett's 1915 work, and Wikipedia. For a detailed narrative of both the French and American efforts—and the broader imperial ambitions behind the canal, see Parker (2007).

BENNETT, I. E. (1915). History of the Panama Canal : its construction and builders. Washington, Historical Publishing Co., 543 pp.

PARKER, M. (2007). Panama Fever: The Epic Story of One of the Greatest Human Achievements of All Time. Anchor Books.

BY THE NUMBERS

1

Brooklyn Bridge *(USA, 1883)*

Achievement: Longest suspension bridge in the world at the time.
Main span: 486 meters.
Previous record: John A. Roebling Suspension Bridge (USA, 1867 – 322 meters).
Difference: ~50% longer.
Notable first: First steel-wire suspension bridge, the tallest towers ever built for a bridge at the time.
Human cost: At least 27 workers died during construction.

2

Vajont Dam *(Italy, 1960)*

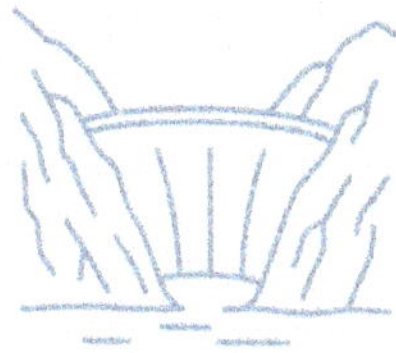

Achievement: Tallest dam in the world.
Height: 261.6 meters.
Previous record: Mauvoisin Dam (Switzerland, 1957 – 237 meters).
Difference: ~10% taller.
Notable risk: Built in a geologically unstable valley.
Tragic aftermath: A massive landslide in 1963 caused a wave that killed ~2,000 people.

3

San Roque Dam *(Argentina, 1891)*

Achievement: Largest artificial reservoir in the world at the time.
Capacity: 260 million cubic meters.
Previous record: Quaker Bridge / San Mateo Reservoirs (USA – 144 million m³).
Difference: ~80% larger.
Legacy: Supplied Cordoba for more than 50 years. Rebuilt, still in use today.

4

Goldfields Water Supply Scheme *(Australia, 1903)*

Achievement: Longest steel pipeline of its time.
Length: 566 kilometers.
Previous record: No predecessor at this scale.
Difference: Record-setting in both length and ambition.
Notable engineering: 60,000 pipes, 8 pumping stations, elevation gain: over 400 meters across arid terrain.
Legacy: Pipeline completed posthumously; still operational today.

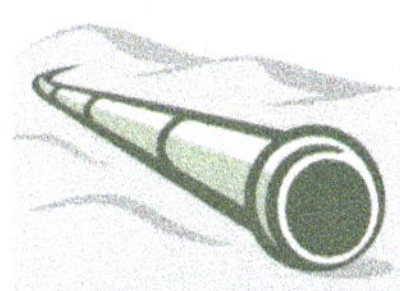

5

Panama Canal *(Panama, 1914)*

Achievement: First interoceanic canal.
Length: 77 kilometers.
Previous record: None – First successful canal to link two oceans.
Human cost: Over 25,000 lives lost during the French phase (1881–1889) due to disease and landslides. Over 5,900 lives lost during the U.S. phase (1904–1914), despite improved sanitation and medical advances.
Project duration: 33 years – 23 years of French effort, followed by 10 years of U.S. direction.

ENDNOTES FROM THE AUTHOR

The Human Spirit Behind Progress

This book began as an exploration of engineering marvels—but it became something far more personal, more human. At the heart of every story lies a single driving force: the human spirit. Each chapter celebrates the will to create, the courage to dream, and the triumph over obstacles once thought impossible. These were not merely feats of engineering—they were acts of vision, of risk, of stubborn hope in the face of doubt, failure, politics, and, at times, death.

Take *Ferdinand de Lesseps*, who—after carving the Suez Canal at the cost of more than 20,000 lives—set out to build another, this time through jungle, mountains, and death. Over 25,000 lives were lost in the process. In today's world, shaped and sometimes obsessed with artificial intelligence (AI), it's worth asking: **Would any AI have approved such a plan?**

Could AI have envisioned a reservoir 80% larger than any before? A dam wall 10% taller? A pipeline eight times longer than anything previously built? Could it have imagined what didn't yet exist—like de Lesseps, O'Connor, Roebling, or Casaffousth once did?

These were not people guided by data, but by a relentless belief that what had never been done could—and must—be done. That kind of visionary defiance—brilliant, reckless, and transformative—remains beyond the reach of AI.

Artificial intelligence excels at handling data, running simulations, and optimizing outcomes. It can calculate the safest bridge span, the most efficient route, and the most economical solution. But it lacks the one thing that drives true innovation: **the ability to imagine beyond what already exists**.

AI, as we conceive it today, is trained on the data of the existing world—and **constrained by what already is**. It favors what is probable, what is safe, and what is predictable. Faced with a radical idea, a well-trained system might respond:

> *"There is no precedent. This is too risky. Too costly. Not feasible."*

A human, instead, might say:

> *"What if?" or "Maybe no one has tried it this way before."*

And yet, it is precisely in **dreaming beyond precedent** that progress is born. Without the human drive to transcend, AI would lead us into a world of perfect repetition—refining the known.

The greatest breakthroughs—the Panama Canal, the Moon landing, the Internet, the printing press—were not the logical outcomes of risk analysis. They were visions. Obsessions. Ideas rejected a hundred times, until someone refused to let go.

AI is a remarkable companion: it sharpens our tools, enhances our decisions, and solves complexity. However, the leap into the unknown—the courage to defy limits not yet imagined—remains a uniquely human trait.

The soul of the greatest innovations will always be ours.

The Hidden Cost

As we dream of the impossible, we must also confront its cost. Behind every structure in this book lie lives lost, landscapes transformed, and immense pressures—personal, social, and political.

The ***Brooklyn Bridge*** demanded the health and lives of its engineers—and of the countless workers who labored and died to make it real. The ***San Roque Dam*** unfolded amidst institutional conflict and sacrifice. The ***Goldfields Pipeline***, though a visionary success, drove Charles Y. O'Connor to the edge.

The ***Panama Canal*** was carved through relentless labor under deadly conditions. And the ***Vajont Dam***—its catastrophe—reveals the danger of ignoring warnings in the pursuit of ambition.

What sacrifices are truly justifiable in the name of progress?

And how do we honor those who bore its cost?

Living on the Edge of Progress

The stories in this book reveal what it means to live—and dream—on the *edge of progress*. The edge is where knowledge meets uncertainty. Where vision encounters resistance. Where ambition walks the razor-thin line between brilliance and tragedy.

Progress lives there—on that edge. It is not driven by those who play it safe, but by those who dare. To move forward, we must imagine further.

These stories have shown that progress demands courage, creativity, and determination. Great advances are not achieved through fear, but through a willingness to explore, to innovate, and to overcome. Those who avoid risk and cling to the familiar rarely chart new directions.

It is this pursuit of the impossible that drives progress.

It means pushing beyond limitations and redefining what's achievable.

But what is progress, truly?

To me, it rests on three pillars: ***Innovation***, which dares to go beyond what is known; ***Social transformation***, which ensures progress improves lives; and ***Sustainability***, which protects the future it seeks to build.

The ***Brooklyn Bridge*** was more than steel and stone—it was a triumph of ingenuity and unity, a symbol of our capacity to overcome both physical and social barriers. The ***San Roque Dam*** spurred the expansion of central Argentina. The ***Goldfields Pipeline*** brought life to the harsh deserts of Western Australia, transforming entire communities and helping the region become a global gold producer. The ***Panama Canal*** rewrote the map of global trade, connecting oceans—and aspirations. The ***Vajont Dam***, tragic and haunting, remains a warning of what happens when ambition outruns caution.

The stories in this book—except for Vajont—stand as testaments to those three pillars in action. They set new standards in construction and design. They reshaped landscapes and enabled societies to grow. And crucially, they did so without compromising what lay ahead.

The future does not belong to those who fear the edge of progress. It belongs to those who choose to navigate it—balancing ambition with responsibility as they redefine what's possible.

This book is a tribute to human ingenuity, effort, and the dreams that push us to the edge of progress.

So as you close this book, consider this:

> ***How far should we go to achieve our dreams?***

The answer may shape the next great stories of progress—**stories that only humanity can write.**

Rocio Luz Fernandez

THE EDGE OF
PROGRESS

Where the impossible begins...

www.ingramcontent.com/pod-product-compliance
Lightning Source LLC
Chambersburg PA
CBHW041748220926
51847CB00013B/141
9786310082820